T0331861

Digital Government and Achieving E–Public Participation:

Emerging Research and Opportunities

Manuel Pedro Rodríguez Bolívar
University of Granada, Spain

María Elicia Cortés Cediel
Universidad Complutense de Madrid, Spain

A volume in the Advances in
Electronic Government, Digital
Divide, and Regional Development
(AEGDDRD) Book Series

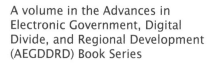

Published in the United States of America by
IGI Global
Information Science Reference (an imprint of IGI Global)
701 E. Chocolate Avenue
Hershey PA, USA 17033
Tel: 717-533-8845
Fax: 717-533-8661
E-mail: cust@igi-global.com
Web site: http://www.igi-global.com

Library of Congress Cataloging-in-Publication Data

Library of Congress Cataloging-in-Publication Data

Names: Rodríguez Bolívar, Manuel Pedro, editor. | Cortés Cediel, María
    Elicia, 1987- editor.
Title: Digital government and achieving E-public participation : emerging
    research and opportunities / Manuel Pedro Rodríguez Bolívar and María
    Elicia Cortés Cediel, editors.
Description: Hershey, PA : Information Science Reference, 2020. | Includes
    bibliographical references and index. | Summary: ""This book examines
    the implications of the implementation of ICTs in the public
    sector"--Provided by publisher"-- Provided by publisher.
Identifiers: LCCN 2019030355 (print) | LCCN 2019030356 (ebook) | ISBN
    9781799815266 (hardcover) | ISBN 9781799815273 (paperback) | ISBN
    9781799815280 (ebook)
Subjects: LCSH: Internet in public administration. | Public
    administration--Information resources management. | Communication in
    public administration--Technological innovations.
Classification: LCC JF1525.A8 S64 2020 (print) | LCC JF1525.A8 (ebook) |
    DDC 352.3/802854678--dc23
LC record available at https://lccn.loc.gov/2019030355
LC ebook record available at https://lccn.loc.gov/2019030356

This book is published in the IGI Global book series Advances in Electronic Government, Digital Divide, and Regional Development (AEGDDRD) (ISSN: 2326-9103; eISSN: 2326-9111)

British Cataloguing in Publication Data
A Cataloguing in Publication record for this book is available from the British Library.

For electronic access to this publication, please contact: eresources@igi-global.com.

Advances in Electronic Government, Digital Divide, and Regional Development (AEGDDRD) Book Series

ISSN:2326-9103
EISSN:2326-9111

Editor-in-Chief: Zaigham Mahmood, University of Derby, UK & North West University, South Africa

MISSION

The successful use of digital technologies (including social media and mobile technologies) to provide public services and foster economic development has become an objective for governments around the world. The development towards electronic government (or e-government) not only affects the efficiency and effectiveness of public services, but also has the potential to transform the nature of government interactions with its citizens. Current research and practice on the adoption of electronic/digital government and the implementation in organizations around the world aims to emphasize the extensiveness of this growing field.

The **Advances in Electronic Government, Digital Divide & Regional Development (AEGDDRD)** book series aims to publish authored, edited and case books encompassing the current and innovative research and practice discussing all aspects of electronic government development, implementation and adoption as well the effective use of the emerging technologies (including social media and mobile technologies) for a more effective electronic governance (or e-governance).

COVERAGE

- Online Government, E-Government, M-Government
- Adoption of Innovation with Respect to E-Government
- Current Research and Emerging Trends in E-Government Development
- ICT Infrastructure and Adoption for E-Government Provision
- ICT within Government and Public Sectors
- Urban Development, Urban Economy
- Citizens Participation and Adoption of E-Government Provision
- Social Media, Web 2.0, and Mobile Technologies in E-Government
- E-Government in Developing Countries and Technology Adoption
- E-Governance and Use of Technology for Effective Government

IGI Global is currently accepting manuscripts for publication within this series. To submit a proposal for a volume in this series, please contact our Acquisition Editors at Acquisitions@igi-global.com or visit: http://www.igi-global.com/publish/.

Titles in this Series

For a list of additional titles in this series, please visit:
http://www.igi-global.com/book-series/advances-electronic-government-digital-divide/37153

Cases on Electronic Record Management in the ESARBCA Region
Segomotso Masegonyana Keakopa (University of Botswana, Botswana) and Tshepho Lydia Mosweu (University of Botswana, Botswana)
Information Science Reference • © 2020 • 300pp • H/C (ISBN: 9781799825272) • US $195.00

Digital Transformation and Its Role in Progressing the Relationship Between States and Their Citizens
Sam B. Edwards III (Quinnipiac University, USA) and Diogo Santos (Pitágoras College, Brazil & Estácio de Sa University, Brazil)
Information Science Reference • © 2020 • 231pp • H/C (ISBN: 9781799831525) • US $185.00

Employing Recent Technologies for Improved Digital Governance
Vasaki Ponnusamy (Universiti Tunku Abdul Rahman, Malaysia) Khalid Rafique (Azad Jammu and Kashmir Information Technology Board, Pakistan) and Noor Zaman (Taylor's University, Malaysia)
Information Science Reference • © 2020 • 383pp • H/C (ISBN: 9781799818519) • US $195.00

Political Decision-Making and Security Intelligence Recent Techniques and Technological Developments
Luisa Dall'Acqua (University of Bologna, Italy & LS TCO, Italy) and Irene M. Gironacci (Swinburne University of Technology, Australia)
Information Science Reference • © 2020 • 169pp • H/C (ISBN: 9781799815624) • US $195.00

For an entire list of titles in this series, please visit:
http://www.igi-global.com/book-series/advances-electronic-government-digital-divide/37153

701 East Chocolate Avenue, Hershey, PA 17033, USA
Tel: 717-533-8845 x100 • Fax: 717-533-8661
E-Mail: cust@igi-global.com • www.igi-global.com

Table of Contents

Preface

The implementation of new technologies (ICTs) in public administrations has introduced relevant public management reforms and changes in civil society (Dunleavy et al., 2005), creating new ways of public and democratic value (Cortés-Cediel et al., 2019). In particular, it has changed how governments work and interact with citizens, who have more accessibility to information and services through the use of technologies, among other aspects (Dixon, 2010; Singh, 2015). But, the most important here is not only the implementation of higher efficiency, transparency and accountability in back-office processes, but also has emphasized the need of introducing new governance models based on a higher citizen engagement in public decisions (Rodríguez Bolívar, 2018). Nonetheless, the implementation of ICT in public administrations has followed heterogeneous patterns (Dixon, 2010; Moon, 2002) that Charalabidis et al. (2019) have categorized into three generations, according to the evolution of e-government.

A first generation in the use of ICTs by governments has involved the development of e-government applications oriented to management and service provision with the aim at increasing efficiency and effectiveness (Charalabidis et al., 2019; Dixon 2010; Alcaide et al., 2014). In the field of service delivery (e-services), e-government has entailed reductions of costs in transactions, changed on the way some public services are delivered, and increased the levels of efficiency in public services delivery. So, e-government has optimized public and human resources in transactions and management activities through the huge impact of technological tools on the promotion of synergies between stakeholders (Rowley, 2010).

Thanks to technological infrastructure in public administrations, these synergies have given rise to e-governance mechanisms promoting Government-to-Citizens (G2C), Government-to-Business (G2B), and Government-to-Government (G2G) relationships (Rowley, 2010; Cortés-Cediel et al., 2017). Government-to-Citizens (G2C) has facilitated the citizens' access to both a variety of relevant information and the performance of transactions, such as tax payments and other bureaucratic tasks, at a lower cost. Government-to-Businesses (G2B) has promoted better environments for innovation and collaboration through higher interaction with corporate bodies and organizations of the private sector, whereas Government-to-Government (G2G) synergies facilitate the interaction between different public agencies, organizations, departments and authorities with the aim at achieving cost reduction and interoperability purposes (Tambouris et al., 2009). In spite of these advantages, the main limitation of this first generation was the cybersecurity due to the vulnerabilities and risks that ICTs produce and the inability of legislation to regulate these risks in an efficient way.

The second generation in the implementation of e-government by public administrations is related to the emergence of open government models (Charalabidis et al., 2019). Specifically, open government has meant the opening of the administration to the citizens, in terms of transparency and collaboration with other stakeholders. Under this context, prior research has mainly focused their efforts on analyzing the technological tools by both governments and citizens, for promoting citizen engagement. According to prior research, web 2.0 technologies are the main tools used in this stage for improving service delivery, responsibility, accountability, and citizen participation (Dixon, 2010; Linders, 2012; Subirats, 2013). Web 2.0 (also known as social media) comprises a series of platform networks that cover all connected devices, allowing users both to collaborate in terms of creating, organizing, connecting and sharing content (Chang & Kanan, 2008; O´Reilly, 2007; Picazo-Vela et al., 2012), and to reduce costs in terms of economic, organizational and time terms (Colombo, 2006).

The dynamics of social media technologies (Alarabiat et al., 2017) and the possibility of content exchange with user-generated content (Kaplan & Haenlein, 2010) have changed the way of participating in terms of spontaneity, immediacy and creation of public value (Colombo, 2006; Salim & Haque, 2015; Rodríguez Bolívar, 2017). Either by the use of ad hoc platforms or by the use of social media, ICTs have also allowed an interconnected society to be resilient in cases of war and natural disasters, such as floods, hurricanes and earthquakes, through mobilization and social activism (Anttiroiko, 2016; Scholl 2019).

Finally, the third generation in the implementation of e-government consists of the incorporation of ICTs in the field of public management that are focused on production, management, and custody of data (Charalabidis et al., 2019). Specifically, according to different authors, data-driven governments are based on: a) the gathering of data through technologies such as IoT devices, social media, cloud computing, and recent technology infrastructures for secure data management like blockchain, and b) data analytics to support policy-making and decision-making (Janssen et al., 2019; Rodríguez Bolívar & Scholl, 2019). This new generation of e-government is focused on both technology-driven processes for automated decision-making and citizen-driven processes. In particular, it has promoted the creation of spaces, such as hackathons, where citizens do not only create data through their electronic devices, but also acquire a leading role in data management.

Analyzing the different generations and phases e-government has experienced highlights the role that the figure of the citizen has acquired in the sphere of decision making. It is not a coincidence that citizens have gained a presence in different areas of public management at the same time that technology has evolved. However, the fact that a public administration is endowed with technological resources does not imply public, social and democratic learning and value achieved through citizen participation in government procedures. In this sense, it is necessary to analyze how the mechanisms for e-participation are being transferred by governments.

E-PARTICIPATION: CITIZEN AS A CENTRAL ACTOR IN THE CREATION OF PUBLIC AND DEMOCRATIC VALUE THROUGH TECHNOLOGY

As noted previously, the implementation of ICTs in public administrations has transformed the way governments manage resources and provide services, especially on how public administration has opened up to the collaboration with other stakeholders (Meijer and Rodríguez Bolívar, 2015), such as civic and economic actors (Bonsón et al., 2012). In this regard, the ICT implementation in governments has pushed the introduction of new governance models in which the citizen emerges as a central actor. Citizens can, therefore, use technological tools both as a complement to the procedures and techniques of representative democracy and as new interaction channels for the generation of new forms of citizenship on its way to a new direct, horizontal and participatory democracy (Medaglia, 2012; León Castro, 2016).

This redistribution of power has emerged in societies through the creation of networks, like those created by the Web 2.0 technologies (Castells, 1996, 2000; Foth et al., 2007; Rodríguez Bolívar, 2015a), and has made different stakeholders to have gained prominence in decision and policy making scenarios (Meijer, 2016). The ultimate aim of this networked society is to enhance the capacity of citizens to actively participate in public decisions, promoting the creation of public value (Rodríguez Bolívar, 2018a and 2018b), increasing the citizens' quality of life (Rodríguez Bolívar, 2019). Under this framework, this book seeks to gather interesting insights regarding worldwide empirical experiences in citizen-centric services and management.

Under this framework, the citizen engagement in public decisions has been told the main issue in the new governance models to make governments more open and close to the citizenry needs (Rodríguez Bolívar, 2016). This way, governments using ubiquitous computing (Salim and Haque, 2015; Weiser, 1999) have enabled mechanisms at different levels of participation that help involving citizens in decision-making from access to institutional information to dynamics of cooperation and co-production (Cortés-Cediel et al., 2019). It has allowed citizens to reduce participation costs and expand information with the aim of taking better decisions (Medaglia, 2012). In particular, the use of e-participation tools established ad hoc or others generalized such as social media has generated a connected society that is able to take the initiative in different participatory, activism and decision-making procedures (Medaglia, 2012; Rotman et al., 2011; Suárez, 2006). This way of procedure is a sign of an interest of a higher-cultural citizenry to participate in different deliberative contexts (Castelnovo et al., 2015; Inglehart, 1991).

All these social and participatory changes have had an impact on democratic theory itself (Albert and Passmore, 2008), which puts the focus in terms of transparency and accountability as objectives of democratic strengthening in increasingly complex societies. Therefore, e-participation influences the way of understanding democracy obtaining higher consensus and a better quality of life in a social sense (Dameri, 2014). Due to the effects that e-participation has had not only on societies, but on the configuration itself and on democracy, governments cannot neglect the therapy that these practices imply in the levels of disaffection that can be given to institutions in political systems. In this context, it is essential to focus on how this is being transferred to the public administration, and analyze the impact and implementation of participation mechanisms by governments, with the aim of improving citizen participation by implementing public policies and providing public services (Giffinger et al., 2007).

SMART CITIES AS SCENARIOS OF CHANGE IN PUBLIC MANAGEMENT

Although difficult to define, the smart city conceptualizations have found consensus on the idea of increasing efficiency and sustainability in a city context with the aim of improving the quality of life of citizens and increasing public value (Caragliu et al., 2009; Rodríguez Bolívar, 2019). In any case, a smart city model understands a multi-dimensional spectrum of urban issues in which different levels of ICTs are used to enhance collective intelligence (Anttiroiko, 2016; Giffinger et al., 2007). Due to the interpretation of cities as spaces formed by human/social and technological infrastructures, smart city models have turned out to be ideal scenarios where the dynamics of e-government and e-governance have evolved. Moreover, they have acquired characteristics of the "smart" concept, resulting in the proliferation of new concepts, such as smart government and smart governance (Meijer & Rodríguez Bolívar, 2015) in which different stakeholders are engaged in the decision-making arena, improving interactions among them (Alawadhi et al., 2012; Brynskov et al., 2014; Meijer & Rodríguez Bolivar, 2015; Willems et al., 2017).

These new forms of collaboration create innovative environments where, through the co-creation of public services by using technology (Rodríguez Bolívar, 2018a), the citizens' quality of life has increased (Albino et al., 2015; Rodríguez Bolívar, 2019). In this context, creative citizens are considered as a resource that contributes value by sharing experience and non-technical knowledge that can be very useful for solving complex problems (Ahlers et al., 2016; Bull & Azenoud, 2016; Rodríguez Bolívar, 2018a), and make it is necessary for governments to take into consideration citizen engagement measures within the governance of a smart city (Castelnovo et al., 2015).

With the aim of not neglecting the value that citizen participation brings, in smart cities initiatives and technologies, such as social media, that seek to enhance citizen engagement are being promoted (Castelnovo et., 2015). Hence, cooperation and co-production initiatives such as crowdsourcing seek to involve citizens more directly and effectively through internet-based applications (Salim & Haque, 2015). In addition to cooperation and co-production, the literature focuses on various disruptive applications of technology in the public management field. This way, recent technologies, such as blockchain, are considered as promising due to their potential in terms of security and transparency (Rodríguez Bolívar & Scholl, 2019)

Due to this heterogeneous panorama between governments, it is useful for technicians, experts and practitioners to have a review of experiences implemented in different parts of the world in order to inspire other possible models of both smart cities, in particular, and e-government, in general.

E-GOVERNMENT EXPERIENCES IN DIGITAL GOVERNMENTS: THE CHOICE OF CHAPTERS

Due to the effects that e-participation has on modern societies, it would be interesting to know how governments are taking steps for achieving a more democratic and open society using ICTs. Different authors have indicated that the implementation of e-government models has different speed and intensity in each public system (Dunleavy et al., 2005). This may mean that acquiring emerging technologies is not being considered in all cases. For this reason, it is important to attend the causes that prevent a high pace in the implementation of technological tools, not only to ensure and enhance transparency and accountability mechanisms in companies, but also for the increase of public value.

Based on the growing interest in the advances on e-government, e-participation and smart cities inside and outside the academic world (Cortés-Cediel et al., 2019), this book presents recent research about the implementation of ICTs in the public sector through different experiences, aiming to understand both the strengths and the vulnerabilities that the management models can entail as well as to contribute to their improvement.

Therefore, this book is divided into three different sections. In the first one, this book collects chapters in which e-government is analyzed from the public management point of view in different governments around the world. The second section of the book collects chapters that are focused on emerging technologies used in different areas of e-participation. The final section of this book collects chapters focused on experiences aimed to address the need of social digitization to empower citizens not only technologically but through technological skills in their task of creating public value.

In Chapter 1, Alcaide-Muñoz, Alcaide-Muñoz and Rodríguez Bolívar show a bibliometric study in the field of e-government with the aim of offering an image of the impact on the focus of attention of the scientific community that has had this new management paradigm based on the use of different technologies. The findings of this analysis show predominant trends of interest about e-Government field by academics. Furthermore, the authors underline the focus of attention of the scientific community in smart cities as innovative spaces for increasing cooperation among governments, facilitating co-creation mechanisms, and enhancing quality of life of citizens.

Contrary to this growing interest in the development of e-government models by the academic community, it is not always possible to implement e-government models. Hence, in Chapter 2 Pinterič has focused on the problem of the lack of motivation of governments to make administrative changes that ensure progress in the implementation of technologies within the public sector. This lack of motivation could not only affect governments, but also spread to other social groups based on different circumstances such as the digital or resource gap, causing spaces where inefficiency can affect public management. In order to shed light on this matter, the author analyzes the case of Slovenia using a methodology based on surveys distributed randomly among 100 citizens on the streets of Maribor.

Similarly, Valle-Cruz and Sandoval-Almazan offer in Chapter 3 an analysis of the adoption of emerging technologies by Mexican state governments. The authors note that although technologies such as cloud computing, big data, Internet of things, and artificial intelligence are emerging, most of Mexican states have a very small advance in the implementation of advances e-government models. On the one hand, the authors indicate the lack of technological infrastructures, including the impossibility of internet connection or electricity in some Mexican regions. On another, it is common for the government to use static web pages and social media as communication mechanisms between government and citizens, instead of other more modern and emerging technologies that provides higher capacities for citizen engagement in public decisions.

The special attention to analyze the technologies used by governments for public management leads Tsabedze to focus his attention on the field of economic transactions in Chapter 4. Specifically, Tsabedze analyzes the management of electronic records (e-records) in Eswatini government. Despite faster communication through the use of e-applications to access government services, the real situation is that the level of e-records readiness in the government ministries is at a preliminary stage. Concretely, Tsabedze indicates that e-records management is poorly handled and he possess the need to advance in this area with the aim at improving the quality of life of citizens. As a contribution to this chapter, Tsabedze makes recommendations on how management of e-records could be improved in the government ministries in Eswatini.

Despite of this, areas such as e-participation arouse interest in governments by articulating dynamics that involve citizens. The E-Participation Index (EPI) developed by the United Nations Department of Economic and Social Affairs shows the quality of political institutions in promoting e-participation. This way, in Chapter 5 Tavares, Martins and Lameiras base their research on data from this index to identify which countries have the best democratic performance. In addition, the analysis sheds light on which administrative levels most promote citizen participation and what are the contextual factors that explain that participation, among other issues.

Chapter 6 addresses the emergence and impact of data as a new raw material of interest to the public sector. For this reason, McNutt and Goldkind analyze the potential of the data not only for the analysis of complex realities but as a solution mechanism for some truly wicket problems. Thus, the use of already existing data known only to experts now extends to other actors that do not belong to the public sector. Specifically, citizens who are volunteer data scientists participate in initiatives address social problems using advanced analytics and large datasets. This is the case of initiatives such as Data for Good, analyzed in this chapter.

Chapter 7 focuses on the analysis of social media as web 2.0-based technologies used by governments. In this way, Hatipoğlu, Zahid Sobaci, and Fürkan Korkmaz analyze the behavior of Turkish public administrations to generate citizen engagement through social networks. According to authors almost 97% of majors in Turkey have a Twitter account. Despite these figures, the presence in the networks by the governments does not imply that social media is used with the aim of generating participation. It is a challenge for governments that social networks are not conceived as mere communication channels but that synergies with citizens create public value.

Other applications based on emerging technologies that are analyzed in this book are those using crowdsourcing. Specifically, in Chapter 8, Yavud, Karkin, and Sevinç Çubuk discuss government crowdsourcing. Authors defined Crowdsourcing as the act of an organization taking a function performed by people outside the organization through an open call online. The authors carry out a study on the literature with the aim of identifying aspects that allow us to propose an integrated model in e-government so that the crowdsourcing dynamics between stakeholders are carried out safely. As revealing findings on this issue, the authors identify organizational aspects that needs to be carefully considered and managed by public managers.

In Chapter 9, Konopacki, Albu, Cerqueira, and Guimarães Tavares focus on an emerging new technological application in Brazil called "Mudamos". This application is initially oriented to sign documents such as bills electronically. However, despite its potential for changing citizen participation, "Mudamos" app became an integrated engagement framework. Therefore, the authors guide the research to identify what factors contribute to getting people involved through the application. To do this, the authors present an engagement framework, connecting cutting-edge digital innovation on electronic signatures with social innovative methodologies with the aim that this kind of tool results a way to create real institutional changes.

In Chapter 10, Muñoz de Luna and Kolotouchkina show the cases of the Smart Cities of London and Madrid as paradigmatic examples in which social and technological infrastructures enhance relationships between stakeholders through participation. Specifically, the authors analyze these cities through practices in the field of digital communication and citizen engagement, identifying different communication channels between different stakeholders. In the findings of this chapter, the authors perceive London and Madrid as a point to the consolidation of a new context for communication and urban management, highlighting the role of citizens as relevant stakeholders to enhance these synergies.

CONCLUSION

The implementation of ICTs in governments has changed the way to understand public management and has had an impact on different domains. Specifically, technology has direct effects on a variety of organizational, behavioral, political and cultural aspects (Dunleavy et al., 2005) and the intensity of these changes directly depends on the level of implementation of the corresponding e-government models. In this regard, the technology has not been used in the same way in all developed countries or even within each government (Dunleavy et al., 2005), which could be due to the different prevalent administrative cultures in each particular case (Cortés-Cediel et al., 2020). As can be seen throughout the chapters of this book, governments have been implementing e-government models according to different needs and priorities, and not all of them have covered the three generations of e-government stages pointed out by Charalabidis et al. (2019).

In this context, according to the technology acceptance model (Lee, 2003), the new generations will have a set of acquired technological competences that will facilitate the reduction of the technological gap. However, although governments are implementing technologies of different nature in order to manage public resources efficiently and sustainably, the truth is that technologies are not always aimed at ensuring that citizens have the capacity to intervene in particular areas and, especially, in decision making processes. In fact, individuals do not only need advanced technological tools, but also the attitudes and skills necessary to handle them, which does not necessary mean that the democratic implication is achieved. Consequently, without specific government strategies, participation will have excessive costs for citizens (Dunleavy et al., 2010; OECD, 2004). A mitigation of this issue will depend on the culture of participation of the society promoted by governments. Governments thus should direct their efforts to advance in the implementation of e-government, and in the pursuit of public value through e-participation mechanisms.

In this sense, this book offers ideas and experiences that illustrate not only proposals and experiences, but also problems and limitations found by governments on their process of implementing e-government models. It is in the face of these challenges that the public administration must promote citizen engagement with the aim of attracting the outputs and outcomes from citizens. For this reason, through the pages of this book, we hope to offer another vision focused on the application of technologies and the limitation they encounter in some government contexts that serves as a guide for public managers and experts in the public sector.

REFERENCES

Ahlers, D., Driscoll, P., Löfström, E., Krogstie, J., & Wyckmans, A. (2016). Understanding Smart Cities as Social machines. *Proceedings of the 25th International Conference Companion on World Wide Web*, 759-764. 10.1145/2872518.2890594

Alarabiat, A., Soares, D. S., & Estevez, E. (2017). *Predicting citizens acceptance of government-led e-participation initiatives through social Media: a theoretical model*. Academic Press.

Alawadhi, S., Aldama-Nalda, A., Chourabi, H., Gil-Garcia, J. R., Leung, S., Mellouli, S., ... Walker, S. (2012). Building understanding of smart city initiatives. *Proceedings of the 2012 International Conference on Electronic Government*.

Albert, A., & Passmore, E. (2008). *Public value and participation: A literature review for the Scottish government*. Scottish Government Research.

Albino, V., Berardi, U., & Dangelico, R. M. (2015). Smart cities: Definitions, dimensions, performance, and initiatives. *Journal of Urban Technology*, *22*(1), 3–21. doi:10.1080/10630732.2014.942092

Alcaide Muñoz, L. A., Rodríguez Bolívar, M. P., & Garde Sánchez, R. (2014). Estudio cienciométrico de la investigación en transparencia informativa, participación ciudadana y prestación de servicios públicos mediante la implementación del e-Gobierno. *Revista de Contabilidad*, *17*(2), 130–142. doi:10.1016/j.rcsar.2014.05.001

Anttiroiko, A. V. (2016). City-as-a-Platform: The Rise of Participatory Innovation Platforms in Finnish Cities. *Sustainability*, *8*(9), 922. doi:10.3390u8090922

Benington, J. (2011). From private choice to public value. Public value. *Theory into Practice*, 31–49.

Bonsón, E., Torres, L., Royo, S., & Flores, F. (2012). Local e-government 2.0: Social media and corporate transparency in municipalities. *Government Information Quarterly*, *29*(2), 123–132. doi:10.1016/j.giq.2011.10.001

Brynskov, M., Bermúdez, J. C. C., Fernández, M., Korsgaard, H., Mulder, I., Piskorek, K., & de Waal, M. (2014). *Urban interaction design: Towards city making*. Academic Press.

Bull, R., & Azennoud, M. (2016). Smart citizens for smart cities: Participating in the future. *Proceedings of Institution of Civil Engineers: Energy*, *169*(3), 93–101. doi:10.1680/jener.15.00030

Caceres, P., Rios, J., De Castro, V., & Insua, D. R. (2007). Improving usability in e-democracy systems: Systematic development of navigation in an e-participatory budget system. *International Journal of Technology, Policy and Management*, *7*(2), 151–166.

Caragliu, A., Del Bo, C., & Nijkamp, P. (2011). Smart Cities in Europe. *Journal of Urban Technology*, *18*(2), 65–82. doi:10.1080/10630732.2011.601117

Castells, M. (1996). *The information age: Economy, society, and culture. Volume I: The rise of the network society*. Academic Press.

Castells, M. (2000). La era de la información: economía, sociedad y cultura. v. 1: La sociedad redInformation age: economy, society and culture. v. 1: The rise of the network society. No. 303.48 C348 2000. Blackwell Publishers.

Castelnovo, W., Misuraca, G., & Savoldelli, A. (2015). Smart Cities Governance: The Need for a Holistic Approach to Assessing Urban Participatory Policy Making. *Social Science Computer Review*.

Castelnovo, W., Misuraca, G., & Savoldelli, A. (2015). Citizen's engagement and value co-production in smart and sustainable cities. *Proceedings of the 2015 International Conference on Public Policy*, 1–16.

Chang, A. M., & Kanan, P. (2014). *Leveraging Web 2.0 in Government*. E-Government/Technology Series, IBM Center for the Business of Government. http://www.businessofgovernment.org/sites/default/files/LeveragingWeb.pdf

Charalabidis, Y., Loukis, E., Alexopoulos, C., & Lachana, Z. (2019, September). The Three Generations of Electronic Government: From Service Provision to Open Data and to Policy Analytics. In *International Conference on Electronic Government* (pp. 3-17). Springer. 10.1007/978-3-030-27325-5_1

Colombo, C. (2006). Innovación democrática y TIC,¿ hacia una democracia participativa? IDP. *Revista de Internet, Derecho y Política*, (3), 28-40.

Cortés-Cediel, M. E., Cantador, I., & Gil, O. (2017). Recommender systems for e-governance in smart cities: State of the art and research opportunities. In *Proceedings of the international workshop on recommender systems for citizens* (pp. 1-6). 10.1145/3127325.3128331

Cortés-Cediel, M. E., Cantador, I., & Rodríguez Bolívar, M. P. (2019). Analyzing Citizen Participation and Engagement in European Smart Cities. *Social Science Computer Review*. doi:10.1177/0894439319877478

Dameri, R. P. (2014). Comparing smart and digital city: Initiatives and strategies in Amsterdam and Genoa. Are they digital and/or smart? In R. P. Dameri & C. Rosenthal-Sabroux (Eds.), *Smart city* (pp. 45–88). Springer.

Dixon, B. E. (2010). Towards e-government 2.0: An assessment of where e-government 2.0 is and where it is headed. *Public Administration and Management*, *15*(2), 418–454.

Dunleavy, P., Margetts, H., Bastow, S., & Tinkler, J. (2006). New public management is dead—Long live digital-era governance. *Journal of Public Administration: Research and Theory*, *16*(3), 467–494. doi:10.1093/jopart/mui057

Foth, M., & Hearn, G. (2007). Networked individualism of urban residents: Discovering the communicative ecology in inner-city apartment buildings. *Information Communication and Society*, *10*(5), 749–772. doi:10.1080/13691180701658095

Giffinger, R. (2007). Smart Cities Ranking of European Medium-sized Cities. *October, 16*, 13–18.

Inglehart, R. (1991). El cambio cultural en las sociedades industriales avanzadas. *CIS, 121*.

Janssen, M., Attard, J., & Alexopoulos, C. (2019, January). Introduction to the Minitrack on data-driven government: creating value from big and open linked data. *Proceedings of the 52nd Hawaii International Conference on System Sciences.* 10.24251/HICSS.2019.349

Kaplan, A. M., & Michael, H. (2010). Users of the world, unite! The challenges and opportunities of social media. *Business Horizons, 53*(1), 59–68. doi:10.1016/j. bushor.2009.09.003

Linders, D. (2012). From e-government to we-government: Defining a typology for citizen coproduction in the age of social media. *Government Information Quarterly, 29*(4), 446–454. doi:10.1016/j.giq.2012.06.003

Medaglia, R. (2012). eParticipation research: Moving characterization forward (2006–2011). *Government Information Quarterly, 29*(3), 346–360. doi:10.1016/j. giq.2012.02.010

Meijer, A. (2016). Coproduction as a structural transformation of the public sector. *International Journal of Public Sector Management, 29*(6), 596–611. doi:10.1108/ IJPSM-01-2016-0001

Meijer, A., & Bolivar, M. P. R. (2015). Governing the smart city: A review of the literature on smart urban governance. *International Review of Administrative Sciences.*

Moon, M. J. (2002). The evolution of e-government among municipalities: Rhetoric or reality? *Public Administration Review, 62*(4), 424–433. doi:10.1111/0033-3352.00196

O'Reilly, T. (2007). What is Web 2.0: Design patterns and business models for the next generation of software. *Communications & Stratégies, 65*, 18–37.

Picazo-Vela, S., Gutiérrez-Martínez, I., & Luna-Reyes, L. F. (2012). Understanding risks, benefits, and strategic alternatives of social media applications in the public sector. *Government Information Quarterly, 29*(4), 504–511. doi:10.1016/j. giq.2012.07.002

Rodríguez Bolívar, M. P. (2015). Policy makers' perceptions on the transformational effect of Web 2.0 technologies on public services delivery. *Electronic Commerce Research, 17*, 1–28.

Rodríguez Bolívar, M. P. (2017). Governance models for the delivery of public services through the Web 2.0 technologies: A political view in large Spanish municipalities. *Social Science Computer Review, 35*(2), 203–225. doi:10.1177/0894439315609919

Rodríguez Bolívar, M. P. (2018a). Creative citizenship: The new wave for collaborative environments in smart cities. *Academia (Caracas), 31,* 1–31.

Rodríguez Bolívar, M. P. (2018b). Governance models and outcomes to foster public value creation in smart cities. *Scienze Regionali, 17*(1), 57–80.

Rodríguez Bolívar, M. P. (2019). In the search for the 'smart' source of the perception of quality of life in European smart cities. *Proceedings of the 52nd Hawaii International Conference on System Sciences,* 3325-3334.

Rodríguez Bolívar, M. P., & Meijer, A. J. (2016). Smart governance: Using a literature review and empirical analysis to build a research model. *Social Science Computer Review, 34*(6), 673–692. doi:10.1177/0894439315611088

Rotman, D., Vieweg, S., Yardi, S., Chi, E., Preece, J., Shneiderman, B., . . . Glaisyer, T. (2011). From slacktivism to activism: participatory culture in the age of social media. In CHI'11 Extended Abstracts on Human Factors in Computing Systems (pp. 819-822). doi:10.1145/1979742.1979543

Rowley, J. (2011). e-Government stakeholders—Who are they and what do they want? *International Journal of Information Management, 31*(1), 53–62. doi:10.1016/j.ijinfomgt.2010.05.005

Salim, F., & Haque, U. (2015). Urban computing in the wild: A survey on large scale participation and citizen engagement with ubiquitous computing, cyber physical systems, and Internet of Things. *International Journal of Human-Computer Studies, 81,* 31–48. doi:10.1016/j.ijhcs.2015.03.003

Scholl, H. J. (2019). Overwhelmed by Brute Force of Nature: First Response Management in the Wake of a Catastrophic Incident. In *International Conference on Electronic Government* (pp. 105-124). Springer. 10.1007/978-3-030-27325-5_9

Scholl, H. J., & Bolívar, M. P. R. (n.d.). Mapping potential impact areas of Blockchain use in the public sector. *Information Polity,* (Preprint), 1-20.

Singh, S. (2015). E-governance state-of-the-art survey: Stuttgart, Germany. In T. M. Vinod Kumar (Ed.), *E-governance for smart cities* (pp. 47–64). Springer.

Suárez, S. L. (2006). Mobile democracy: Text messages, voter turnout and the 2004 Spanish general election. *Representation (McDougall Trust), 42*(2), 117–128. doi:10.1080/00344890600736358

Subirats, J. (2013). Internet y participación política: ¿nueva política? ¿nuevos actores? *Revista de Ciencias Sociales*, *26*(33), 55–72.

Tambouris, E., Loutas, N., Peristeras, V., & Tarabanis, K. (2009). The role of interoperability in egovernment applications: An investigation of critical factors. *Journal of Digital Information Management*, *7*(4), 235–243.

Weiser, M., Gold, R., & Brown, J. S. (1999). The origins of ubiquitous computing research at PARC in the late 1980s. *IBM Systems Journal*, *38*(4), 693–696. doi:10.1147j.384.0693

Willems, J., Van den Bergh, J., & Viaene, S. (2017). Smart City Projects and Citizen Participation: The Case of London. In R. Andessner, D. Greiling, & R. Vogel (Eds.), *Public Sector Management in a Globalized World* (pp. 249–266). Springer Fachmedien Wiesbaden. doi:10.1007/978-3-658-16112-5_12

Chapter 1
Analyzing of the Evolution of the Field of E-Government and Trending Research Topics:
A Bibliometric Study

Laura Alcaide-Muñoz
University of Granada, Spain

Cristina Alcaide-Muñoz
iD https://orcid.org/0000-0001-6910-202X
International University of La Rioja, Spain

Manuel Pedro Rodríguez Bolívar
iD https://orcid.org/0000-0001-8959-7664
University of Granada, Spain

ABSTRACT

e-Government is a research topic that arouses the interest of many researchers all around the world. So, we can find a large number of studies and research projects published about this topic. Given the large number of articles that exist in the literature, it is not possible to get an idea of the evolution shown by the field of study and see the topics that are not receiving attention from researchers. The objective of this chapter is an analysis of the academic literature on e-government and the evolution of this field of knowledge. These findings allow us to have a clear idea of the evolution of e-government field, the disappeared research topics, and those that are currently in a lively debate. This analysis could be of interest to identify the trend in research of the e-government field of knowledge, as well as to examine the specialization of certain research topics.

DOI: 10.4018/978-1-7998-1526-6.ch001

INTRODUCTION

Many governments Many governments around the world have carried on many innovative e-Government projects (Anthopoulos & Fitsillis, 2014), given that these initiatives have promoted the transformation of public management adopting technological advances that favor democratic legitimacy, participation in the configuration of public policies and in the transparency of public resources management. (Aham-Anyanwu & Li, 2017). These initiatives favor access to public information which allows citizen to be informed of the public policies and decisions adopted by public managers and politicians (Karamagioli et al., 2014), increasing trust in governments (Ohemeng & Ofosu-Adarkwa, 2014) and enhancingcitizen participation in public affairs (Ahn & Berardino, 2014; Rodríguez Bolívar, 2015).

In this regard, governments use the social networks more frequently to relationship with citizenship, civil organizations, firms, and so on, spreading information and encouraging citizen participation (Aladalah et al., 2018; Maxwell & Carboni, 2017). Through these networks, citizens can communicate their perceptions, expectations and experiences with public services, and can also demand services that better satisfy their needs (Rodríguez Bolívar & Alcaide Muñoz, 2018).

These increasing demands of the citizenry push governments to develop tools, apps and platforms to facilitate the participation in decision-making in public affairs. This way, the governments have to face these demands, undertaking initiatives and projects involving different stakeholders (Wimmer & Scherer, 2018). The new technologies Web 2.0 facilitate the context and environment to undertake collaborative projects, and those in which citizens can participate in the co-creation of public services achieve more citizen-centric services adapted to their needs.

In addition, the initiatives of Smart Cities (SCs) favor the innovative and technological spaces to promotes the citizen participation, which enables the cooperation and co-creation among governments, organizations and citizens (Ferro & Osella, 2017), which allows to improve the citizenship's quality of life. So, the environment of SCs favors the direct connection with citizenship (Deakin & Reid, 2017), allowing them to solve their own problems with the technology available through e-Government practices or even with technologies created for collaboration with citizenry.

Therefore, e-Government is a research topic that arouses the interest of many researchers all around the world (Alcaide Muñoz et al., 2018). So, we can find a great number of studies and research projects published about this topic (Alcaide Muñoz & Rodríguez Bolívar, 2015). Previous studies have tried to offer information about contextualization of this field of knowledge (Alcaide Muñoz & Garde Sánchez, 2014). However, we go further and offer an improved previous version of e-Government study (Alcaide Muñoz & Garde Sánchez, 2014), offering evolution of this field, the disinterested topics, the trending topic and so on. This analysis could be useful to identify the state of the art in e-Government, offering research possibilities for the future.

Thus, this article aims at assisting researchers to develop e-Government. To achieve this aim, we have analyzed e-Government articles published classified in three sub-periods: pre-crisis (2000-2008), crisis (2009-2013) and post-crisis (2014-2017) periods. Hence, this paper seeks to answer each of the following research questions:

RQ1: How many e-Government articles have been published in the analyzed JCR journals? Will this selection of articles reveal any trend?

RQ2: What research methodology is used in analyzing e-Government? Which countries make the most important contributions in this respect?

RQ3: Which universities and departments make the most important contributions in this respect?

Therefore, the aim of this article is to analyze the academic literature on e-Government and the evolution of this field of knowledge. These findings add new insight of the evolution of e-Government field, the disappeared research topics, and those that are currently in a lively debate. This analysis could be helpful to identify the trend in research of the e-Government field of knowledge, and to examine the specialization of certain research topics.

This article is organized as follow. In Section 2, we review the main bibliometric studies developed in the field of e-Government. Section 3, the research strategy developed in this article is explained, making specific reference to the selection process of the sample and the methodology used and, thereupon, the results obtained from the proposed research. Finally, the article closes with discussion and final remarks.

BIBLIOMETRIC STUDIES IN E-GOVERNMENT RESEARCH

Scholars identify the historical roots of a particular field of study or research topic by bibliometric methodology (Atkins, 1988). It also allows them to predict novel research streams (Löfstedt, 2005) as well as to identify where these studies should be addressed (Webster & Watson, 2002).

Yildiz (2007) highlighted the main limitations of previous research in the e-Government field through bibliometric methods. After a critical evaluation of previous studies of this research topic, such as vagueness in defining this term, he emphasized the need for empirical studies with which to obtain new theoretical arguments, concepts and categories.

Heeks and Bailur (2007), on the other hand, offered an overview of the field of e-Government, and built the corresponding research philosophy. Also, Scholl (2009) focused on the outlook for this related research topics and provided a comprehensive description of the contributions developed by relevant researchers in the field of e-Government, the names of the most prolific researchers, the most commonly studied topics and the main journals and conferences. Therefore, it allows researchers to identify the best media for their publications.

Nonetheless, both researchers (Heeks & Bailur, 2007; Scholl, 2009) present serious limitations because they only focused on specific aspects of e-Government and on a limited number of conferences and journals, ignoring articles published on e-Government in other leading journals, such as Information Society and Social Science Computer Review, as well as others relevant to public administration such as American Review of Public Administration or Public Administration Review, all of which are of high quality and constitute valid research references.

Lastly, Rodríguez Bolívar et al. (2015) pay specific attention to the gaps in knowledge in emerging countries and develop a comprehensive analysis of the past, which can result in steps forwards in future research and findings, stressing the potential for research into e-Government. In addition, Alcaide Muñoz & Rodríguez Bolívar (2015) provide guidelines for researcher who look for direction for future research projects, exploring research trends in e-Government and examined the most used methodologies used in this field.

Other, previous studies have tried to offer information about contextualization of this field of knowledge (Alcaide Muñoz & Garde Sánchez, 2014), but this study has limitations because its results are during a specific stage (2009-2012). Although the conclusions are useful for the researchers, the vision is fragmented and limited to four years. In this paper, we go further and offer an improved previous version of e-Government study (Alcaide Muñoz & Garde Sánchez, 2014), offering evolution of this field (2000-2017), the disinterested topics, the trending topics, and so on. Therefore, the main motivation to develop this empirical study has been the lack of evidence and conclusions in order to reach a greater understanding of the major issues related to e-Government.

BIBLIOMETRIC APPROACH

Sample Selection

It is generally accepted that the journals publications are not only the main source of new knowledge used by academics, but also a means for its disclosure, and an indicator of scientific productivity (Legge & Devore, 1987; Nord & Nord, 1995). In addition, because of the limited view of the subject offered by symposia, summaries o f papers read, letters to the editor, articles of a professional nature a books review, our studies mostly focused on journal publications. However, we also took into consideration articles included in special issues of journals, since they reflect a greater interest in the study of a particular issue and the need to examine it further (Alcaide Muñoz & Rodríguez Bolívar, 2015).

Objective indicators were used to select journals for our study (Forrester & Watson, 1994). We seek references that provided useful and reasonably valid statements in terms of research consumption (Garfield, 1972). Taking into account the findings of previous studies (Alcaide Muñoz & Garde Sánchez, 2015), we excluded journals of marginal importance, i.e. those with an impact factor of less than 0.25 or with fewer than 50 total citations. Finally, we have focused our analysis on the Public Administration and Information Science and Library Science field of knowledge, because more than 70% of the articles on e-Government were published in journals listed in these fields (Rodríguez Bolívar et al., 2014).

Taking into account this requirement, we prepare the list of journals (whole of sample) that were analyzed, for which we take as reference the last impact factor published by Web of Science (WoS) –impact factor 2016-. In this sense, we start with an initial sample of 111 journals, 44 Public Administration journals and 67 Information Science & Library Science journals. But not all of them published e-Government articles –see Table 1-.

To select articles to be analyzed, we have reviewed all the articles in each of the journals that meet the condition described above (Alcaide Muñoz et al., 2014). Firstly, the title, the abstract and the keywords of each one was examined (Lan & Anders, 2000; Plümper & Radaelli, 2004). Afterwards, the introduction was read by the researchers in order to identify the research goals and to determine the main factors analyzed. Finally, if the preceding criteria were insufficient, the papers are read in detail

Taking into account that the economic crisis began to take hold between 2007 and 2008 (Navarro et al., 2016), it seems logical to assume that this type of study would have begun to appear from 2009 onwards, and therefore we have determined three sub-periods, pre-crisis (2000-2008), crisis (2009-2012) and finally, post-crisis (2013-2017). Consequently, our database is composed of 1,332 articles published about e-Government in 64 journals catalogued by the ISI as belonging to the areas of Public Administration (443) and Information Science & Library Science (889), of which 448 articles (2000-2008), 347 articles (2009-2012) and 537 articles (2013-2017) –see Tables 1 and 2-.

Table 1. e-Government articles found in each of the ISI Journals (2000-2017)

Position	Field	Abbreviated journal name	Impact factor 2016	Total articles	Articles of e-government		
					2000-2008	2009-2012	2013-2017
Q1	I.S.	MIS QUART	7.268	2	1	0	1
	I.S.	J INF TECHNOL	6.953	4	0	1	3
	I.S.	INFORM SYST J	4.122	10	6	2	2
	I.S.	J COMPUT-MEDIAT COMM	4.113	18	9	2	7
	I.S.	GOV INFORM Q	4.09	415	98	132	185
	I.S.	INT J INFORM MANAGE	3.872	37	10	9	18
	P.A.	J PUBL ADM RES THEOR	3.624	18	10	4	4
	I.S.	J STRATEGIC INF SYST	3.486	11	7	1	3
	P.A.	PUBLIC ADMIN REV	3.473	52	24	16	12
	I.S.	TELEM INFORM	3.398	14	0	0	14
	I.S.	INFORM & MANAG	3.317	19	9	2	8
	P.A.	PUBLIC ADMIN	2.959	24	11	8	5
	I.S.	EUR J INFORM SYST	2.819	21	9	7	5
	I.S.	INFORM SYST RESEACH	2.763	3	0	1	2
	P.A	GOVERNANCE	2.603	11	8	2	1
	I.S.	INFORM PROCESS MANAG	2.391	2	0	1	1
	I.S.	J MANAGE INFORM SYST	2.356	4	2	2	0
	I.S	J ASSOC INFORM SCI TECH	2.322	9	3	5	1
	I.S.	SOC SCI COMPUT REV	2.293	89	23	22	44
	P.A.	PUBLIC MANAG REV	2.293	22	2	10	10
	P.A.	POLICY STUD J	2.153	3	1	2	0
	I.S.	SCIENTOMETRICS	2.147	5	1	3	1
	P.A.	POLICY AND POLITICS	1.939	5	3	2	0
Q2	I.S.	J ASSOC INF SYST	2.109	6	1	2	3
	P.A.	J PUBLIC POLICY	1.778	7	6	1	0
	P.A.	ENVIRON PLANN C	1.771	4	2	2	0
	P.A.	POLICY SCI	1.75	3	0	2	1
	P.A.	INT PUBLIC MANAG J	1.723	8	1	4	3
	I.S.	INFORM DEV	1.691	25	2	4	19
	P.A.	J EUR SOC POLICY	1593	1	1	0	0
	P.A.	REV POL RESEARCH	1.562	9	6	1	2
	I.S.	INF SOCIETY	1.558	31	18	8	5
	I.S.	ONLINE INFORM REV	1.534	20	6	6	8
	I.S.	TELECOMMUN POLICY	1.526	23	8	4	11
	I.S.	ASLIB J INF MAN	1.514	26	18	2	6
	I.S.	ETHICS INFOR TECH	1.5	2	0	2	0
	P.A.	REV PUB PERSONNEL ADM	1.474	8	4	1	3
	P.A.	J SOC POLICY	1.458	2	2	0	0
	P.A.	AM REV PUBLIC ADM	1.438	48	23	5	20
	I.S.	J INF SCI	1.372	13	9	1	3
	P.A.	INT REV ADM SCI	1.35	39	16	10	13
	I.S.	INFORM TECHNOL PEOPL	1.339	19	6	7	6
	I.S.	INFORM TECHNOL DEV	1.333	32	5	11	16

continued on following page

Table 1. Continued

				Articles of e-government			
Q3	P.A.	SCI PUBLIC POLICY	1.538	5	1	1	3
	P.A.	J ACCOUNT PUBLIC POL	1.333	2	2	0	0
	I.S.	J GLOB INF TECH MANAG	1.167	6	4	2	0
	P.A.	PUBLIC MONEY MANAGE	1.133	9	7	0	2
	P.A.	POLICY AND SOCIETY	1.115	1	0	1	0
	P.A.	ADMIN SOC	1.092	32	11	7	14
	P.A.	AUST J PUBLIC ADMIN	1.072	12	4	5	3
	I.S.	INFORM TECHNOL MANAG	1.067	3	1	2	0
	P.A.	J COMP POLICY ANAL	1.017	4	0	1	3
	P.A.	LOCAL GOV STUD	0.93	28	6	7	15
	P.A.	PUBLIC ADMIN DEVELOP	0.86	19	14	1	4
	I.S.	MALASYAN J LIB INF SCI	0.65	1	1	0	0
	I.S.	INFORM RES	0.574	4	1	2	1
	I.S.	J GLOB INF MANAG	0.517	15	11	1	3
Q4	P.A.	PUBLIC PERFROM MANAG	0.812	9	3	0	6
	P.A.	LEX LOCALIS	0.714	9	0	0	9
	P.A.	POLICY STUD-UK	0.609	3	0	1	2
	P.A.	J HOMELAND SEC EM MANAG	0.474	1	0	1	0
	P.A.	TRANSYLV REV ADM SCI	0.456	22	3	6	13
	P.A.	CAN PUBLIC ADMIN	0.333	8	3	1	4
	P.A.	GESTION Y POLITICA PUBLICA	0.324	15	5	1	9
TOTAL				1332	448	347	537

Sources: Own elaboration with the information from ISI of Knowledge

Abbreviations: P.A. (Public Administration), I.S. (Information Science and Library Science)

NOTE: This table shows only those journals that have published articles about e-Government

Content Analysis

To test our database and, thus, to identify the issued discussed and the methodologies applied in the articles, we use exploratory content analysis (Krippendorff, 1980). Afterwards, we choose QSR NVivo v.11 software to automate the coding of the articles (Fraser, 2000), taking advantage of the option provided to construct random labels, thus obtaining a hierarchized structure of concepts.

During encoding phase, several meeting was held to decide the labels to be assigned and the topics to be included (see Table 2). Thereupon, we encoded each of the articles included in the study sample individually (Lan & Anders, 2000), and we discussed and solved any disagreements relating to the definition of the categories to be analyzed.

After cataloguing and systematizing the specific areas under study relating to e-government, the methodologies applied in each line of research were examined to identify the research trends present. To avoided double counting related to methodologies used in an article, we focuses on the main research goal and on the methodology incorporated to achieve it. So, it was critical to recognize the major intention of the scientific document.

Table 2. Chronological distribution of e-Government research topics

	2000-2008				2009-2012				2013-2017			
	P.A.	I.S.	% P.A.	% I.S.	P.A.	I.S.	% P.A.	% I.S.	P.A.	I.S.	% P.A.	% I.S.
Technological innovation	28	40	15.64%	14.87%	19	36	18.45%	14.75%	25	62	15.53%	16.49%
E-Participation and Web 2.0	45	45	25.14%	16.73%	34	58	33.01%	23.77%	61	98	37.89%	26.06%
Delivery and public services	24	38	13.41%	14.13%	13	44	12.62%	18.03%	18	57	11.18%	15.16%
Governmental transparency	22	36	12.29%	13.38%	16	21	15.53%	8.61%	37	52	22.98%	13.83%
Role of public-sector workers	8	8	4.47%	2.97%	2	4	1.94%	1.64%	4	6	2.48%	1.60%
Legislative architecture	5	10	2.79%	3.72%	0	6	0.00%	2.46%	0	6	0.00%	1.60%
Interoperatibility	4	12	2.23%	4.46%	5	16	4.85%	6.56%	1	6	0.62%	1.60%
Digital divide and resistance to change	2	16	1.12%	5.95%	2	16	1.94%	6.56%	2	15	1.24%	3.99%
Organizational and institutional change	10	25	5.59%	9.29%	7	15	6.80%	6.15%	3	28	1.86%	7.45%
Evalutation and analysis of public policies	31	39	17.32%	14.50%	5	28	4.85%	11.48%	5	20	3.11%	5.32%
Smart cities	0	0	0.00%	0.00%	0	0	0.00%	0.00%	5	26	3.11%	6.91%
Total	179	269	100%	100%	103	244	100%	100%	161	376	100%	100%

Sources: The authors
Abbreviations: P.A. (Public Administration), I.S. (Information Science and Library Science)

ANALYSIS OF RESULTS

RQ1: How many e-Government articles have been published in JCR journals? Will this selection of articles reveal any trend?

Although, previous studies have shown a gradual increase in studies carried out on e-Government over last years (Alcaide Muñoz & Rodríguez Bolívar, 2015; Rodríguez Bolívar et al., 2016), the achieved results in this study show that the number of studies have increased in the post-crisis period, exist for the years 2014 and 2015 –see Figure 1-. Even so, the published articles offer a growing trend, which reflect the continuing interest of researchers in the e-Government field of knowledge.

In this context, the most of analyzed subjects in terms of e-Government are, firstly, the adoption of technological advances to foster the citizens' participation in public

Figure 1. Chronological distribution of e-Government articles found in each of the ISI Journals (2000-2017)
Sources: The authors

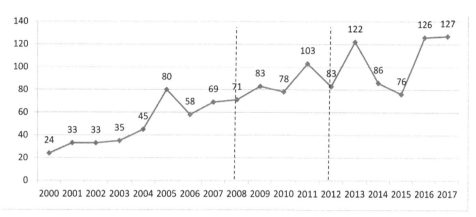

management (25.45%; 339/1,332). This topic shows a growing trend over years – see Figure 2-, with special increased in post-crisis period in Public Administration journals –see Table 2-. Secondly, articles about the impact ICT on the modernization of governments (15.77%; 210/1,332) offered a decrease in the number of published articles in crisis-period to go up back in post-crisis period in both areas of knowledge –see Figure 2-. Thirdly, the study emphasizes how the adoption of e-Government boosts productivity in public services (14.56%; 194/1332) increasing the satisfaction of citizens. These studies have increased in number of published studies, but have reduced the importance in both areas of knowledge in post-crisis periods –see Figure 2-. Finally, governmental transparency and accountability is a research topic that has increased its appearance over years, especially in post-crisis period in Public Administration journals –see Figure 2-. According to Alcaide Muñoz et al., (2017), all of these research topics are well development and important for the structuring of the e-Government field. Similarly, our results show that the appearance of a new topic (Smart Cities) in post-crisis period, which has overtaken other topics like as organizational and institutional change, digital divide or analysis of public policies.

RQ2: What research methodology is employed in analyzing e-Government? Which countries make the most important contributions in this respect?

As for the methodology used in the articles published on the topics analyzed, the researchers that analysis e-Government are tended to use empirical research methods (90.54%; 1206/1332 versus non-empirical techniques 9.46%; 126/1332) –see Table 3-. Likewise, the results show a clear trend for the use of qualitative methodologies

Figure 2. Main e-Government topics addressed in the leading journals in the fields of public administration and information science
Sources: The author

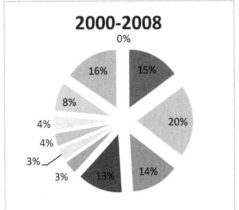
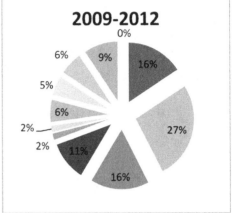
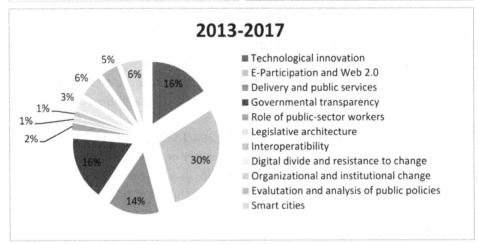

over years in pre-crisis and crisis periods, but this preference is not so obvious in post-crisis periods. The most recent studies are more prone to use qualitative and quantitative methodologies, instead of non-empirical methods for the analysis of e-government phenomena. In addition, the use of quantitative methodologies by the researchers has increased over years, reaching values of 46.58% and 46.01% in Public Administration and Information Science, respectively. This new tendency suggests that e-Government is a research field that over time has gradually acquiring a certain maturity, since the academic researchers are trying in their latest studies to test empirically the practical confirmation of previously defined theories.

Table 3. Chronological distribution of the use of methodologies

YEARS	FIELD	NON-EMPIRICAL	QUALITATIVE METHODOLOGIES	QUANTITATIVE METHODOLOGIES	TOTAL	% NON-EMPIRICAL	% QUALITATIVE METHODOLOGIES	% QUANTITATIVE METHDOLOGIES
2000-2008	P.A.	29	109	41	179	16.20%	60.89%	22.91%
	I.S	32	170	67	269	11.90%	63.20%	24.91%
2009-2012	P.A.	13	55	35	103	12.62%	53.40%	33.98%
	I.S	18	146	80	244	7.38%	59.84%	32.79%
2013-2017	P.A.	14	72	75	161	8.70%	44.72%	46.58%
	I.S	20	183	173	376	5.32%	48.67%	46.01%
	TOTAL	126	735	471	1332	9.46%	55.18%	35.36%

Sources: The authors

Abbreviations: P.A. (Public Administration), I.S. (Information Science and Library Science)

As for the universities that investigate on e-Government, the majority of published papers come from European, USA and Canadian Universities –see Figure 3-. In this sense, we can observe that the impact of European (Western and Eastern) universities has increased over years, specially, in the post-crisis period with 48.42% and 29.89% respectively – see Figure 3-. These Universities of European Countries have increased their interest in E-Participation, in the use of Social Media and Gov2.0, and in the improvement of government transparency and accountability.

In the case of USA's universities, research has suffered a decrease during analyzed the global period, with an increase of studies about e-Participation and governmental transparency, and a decrease of public services research. Canadian universities have also decreased their participation in the studies published on e-Government. This reduction was greater in times of crisis, although in post-crisis period their research increases – see Figure 3-. The greatest increase has been focused on studies about implementation of initiatives of e-Government and access to information.

Regarding Latin American universities, we can observe that their participation in published articles show a growing trend over years. These universities offer an increase in articles about implementation of e-Government and Smart Cities, although they have decreased research projects about analyzing of a-participation channels.

On the other hand, the Asian universities also have suffered a growing trend over years, especially, the main increases have occurred in articles about the adoption of e-Government, e-Participation, public services and Smart Cities – see Figure 3-. Finally, the Australian universities and New Zealand universities have kept their scientific production at present. Their researchers show special interest in the adoption of e-Government, use of Social Media or others tools that favor the citizens' participation, government transparency, and Smart Cities (specially, in the case of New Zealand universities).

RQ3: Which universities and departments make the most important contributions in this respect?

Table 4 shows the evolution of the published articles by the departments. The departments that have increased their publication in the e-Government are Public Administration, Computer Science, Public and Political Science, Management and Information Systems and Accounting (in this order). In the case of Public Administration and Management departments, they show a similar behavior. Their researchers are interested in the adoption of e-Government, e-Participation, Public services and Smart Cities, and this has increased over time.

Public and Political Science and Accounting departments have similar interest and analysis. Their academics have increased their studies about e-Participation issues and transparency. Finally, the Computer Science departments have increased their analysis about the use of Gov2.0, social media and other e-Participation tools, and the platforms and structure of public services.

CONCLUSION

The achieved findings of this study highlight that the prominent research topics are the use of technological advances to promote the citizens' participation in public management, how to adopt of new technologies impact on the transformation of organizational structural of governments and the adoption of e-Government boosts productivity in public services. Also, governmental transparency and accountability is research that has been always analyzed but in post-crisis period their published papers have increased.

Likewise, these findings evidence that the studies about public services analyzed the online public services delivery and the revolution that led to the adoption of new technologies. The Smart Cities' context promotes the innovative spaces to increase the cooperation among governments, citizen and organization, and facilitate the co-creation the better public services, which increase the quality of life of citizens.

Multiple aspects of the E-government field have not been adequately examined by specialists, for instance the issues concerning the varying barriers and restrictions faced by citizens, specially, when they try to get access to more interactive tools, that leads to partial participation or how the adopted governmental strategies may foster this participation (Alcaide Muñoz et al., 2017). Moreover, there is a need to provide different ways of effective citizen participation in order to build ideas and promote novel initiatives on public online services. Future research should focus on how Living Labs favor the development of innovative ideas and solutions.

Figure 3. Main e-Government topics and countries addressed in the leading journals in the fields of public administration and information science
Sources: The authors. Abbreviations: Techn. Innov. (Technological innovation), E-Part. Web 2.0. (E-participation and Web 2.0), Del. Pub. Serv. (Delivery and public services), Govern. Transp. (Governmental transparency), Rol. Pub. Workers (Role of public-sector workers), Leg. Archit. (Legislative architecture), Interop. (Interoperatibility), D. D. Res. to Chan. (Delivery divide and resistance to change), Org. Inst Chan. (Organizational and institutional change), E. Anal. Pub. Pol. (Evaluation and analysis of public policies

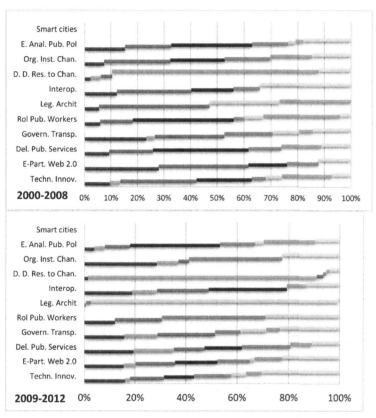

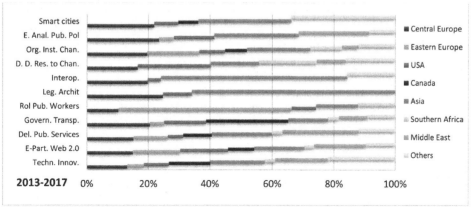

Table 4. Chronological distribution of e-Government research topics and departments

2000-2008	Account and Econ.		Mark. and Comm.		Computer Science		Public Admin.		Management and Business		Public and Political Sc.		Library and Inform. Sc.		Engineering		Others	
	P.A	I.S	P.A	I.S	P.A	I.S	P.A	I.S	P.A	I.S	P.A	I.S	P.A	I.S	P.A	I.S	P.A	I.S
Technological innovation	1	1	2	6	0	23	18	9	5	13	11	7	0	5	2	3	2	6
E-Participation and Web 2.0	1	1	0	21	1	20	28	11	7	6	27	7	0	7	0	0	16	4
Delivery and public services	3	6	5	9	2	22	12	4	7	19	8	6	0	11	1	0	3	5
Governmental transparency	18	8	0	9	2	3	10	5	3	10	6	4	0	10	3	0	3	8
Role of public-sector workers	0	1	0	4	2	1	5	0	1	4	5	6	0	0	0	1	0	1
Legislative architecture	0	0	0	1	0	1	4	0	0	1	1	1	0	0	0	0	2	3
Interoperatibility	0	0	0	4	1	6	3	14	0	1	3	0	0	2	0	0	0	0
Dig. divide and resist. to change	0	4	0	4	0	3	1	4	0	6	0	0	0	3	0	0	1	4
Org. and inst. change	0	2	1	3	0	23	12	7	0	11	2	0	0	4	0	2	0	1
Eval. and analysis of pub. pol.	6	6	0	4	6	37	14	6	7	6	14	4	0	9	0	1	5	4
Smart cities	0	0	0	0	0	0	0	0	0	0	0	0	0	0	0	0	0	0
Total	29	29	8	65	14	139	107	60	30	77	77	35	0	51	6	7	32	36
2009-2012																		
Technological innovation	4	10	2	3	2	40	14	6	7	18	8	2	0	3	0	0	5	5
E-Participation and Web 2.0	0	11	0	24	2	45	40	11	3	14	10	8	0	10	0	2	9	17
Delivery and public services	8	8	5	10	1	38	10	9	4	37	3	4	0	2	0	0	5	6
Governmental transparency	14	7	0	8	0	2	14	4	0	7	3	3	0	8	0	2	3	1
Role of public-sector workers	0	0	0	0	2	6	1	0	0	2	0	1	0	0	0	0	0	0
Legislative architecture	0	1	0	4	0	1	0	1	0	4	0	0	0	0	0	0	0	1
Interoperatibility	0	5	0	1	0	17	8	4	0	5	1	0	0	6	0	0	1	2
Dig. divide and resist. to change	0	1	0	2	0	12	0	1	0	9	1	3	0	0	0	0	0	4
Org. and inst. change	0	0	1	1	0	15	6	6	2	4	2	0	0	4	0	0	0	0
Eval. and analysis of pub. pol.	0	0	0	8	1	27	4	3	0	12	2	1	0	10	0	1	0	1
Smart cities	0	0	0	0	0	0	0	0	0	0	0	0	0	0	0	0	0	0
Total	26	43	8	61	8	203	97	45	16	112	30	22	0	43	0	5	23	37
2013-2017																		
Technological innovation	3	8	1	2	2	49	26	20	1	41	3	11	0	9	2	5	2	8
E-Participation and Web 2.0	18	21	5	53	2	50	46	28	12	27	28	31	0	6	3	2	8	16
Delivery and public services	6	10	0	1	6	56	23	11	4	44	3	9	0	0	0	4	1	7
Governmental transparency	43	20	0	10	0	35	14	23	1	21	23	10	0	4	0	2	3	6
Role of public-sector workers	1	0	0	2	0	6	3	0	3	1	1	0	0	2	0	3	0	2
Legislative architecture	0	0	0	1	0	5	0	0	0	1	0	0	0	2	0	0	0	0
Interoperatibility	0	1	0	0	0	5	0	7	0	1	1	1	0	1	0	0	0	0
Dig. divide and resist. to change	1	0	0	7	0	14	0	5	0	6	1	1	0	1	0	0	5	4
Org. and inst. change	0	1	0	0	0	23	3	12	2	22	0	6	0	0	0	0	2	2
Eval. and analysis of pub. pol.	5	0	0	7	0	24	3	7	0	8	3	1	0	1	0	1	0	0
Smart cities	1	5	0	4	0	19	6	21	1	12	0	3	0	3	0	0	5	8
Total	78	66	6	87	10	286	124	134	24	184	63	73	0	29	5	17	26	53

Sources: The authors

Abbreviations: P.A. (Public Administration), I.S. (Information science), Account. and Econ. (Accounting and Economics); Mark. and Comm. (Marketing and Communication), Public Admin. (Public Administration), Public and Political Sc. (Public and Political Science), Library and Inform. Sc. (Library and Information Science).

In addition, there is still scope for development and improvement in this field, and it is needed to reach a greater understanding of understand the perceptions and attitudes of public managers and political leaders (Norris & Reddick, 2013). Also, it is essential to analyze the planning and decision-making processes (strategic projects) developed in order to better understand the complexity inherent in the introduction of innovation in the public sector. In this context, it is needed studies that analyzed the role of public managers and strategic tools used in the decision-making process.

As for methodology applied, there is a prominent use of empirical methodology in the field of e-Government, with a particular interest in a quantitative approach. Hence, the primary methodologies are the regression analysis, factory analysis, structural equation model, and so on.

Nonetheless, a slight change in the methodological tools applied has been identified. In this sense, quantitative methods are increasingly used by researchers in recent years, and we can also observe that a large of varied methodologies is applied. Thus, we can claim that e-Government is an eclectic subject field which involves several different academic disciplines and research fields. Each research field links its own theories and methodologies to the subject and uses the methods and techniques that it considers appropriate to analyze this phenomenon.

It is also clear that it has been researchers from European, USA and Canadian Universities who have mainly contributed to this field. Particularly, the most of their research activities focuses on the study of specific cases of the application of new technology in public administrations. This has been the case for two reasons: on the one hand, because these researchers are interested in e-Participation, in the use of Social Media and Gov 2.0, and the improvement of government, transparency and accountability.

It will be needed to know the successful experience in Smart Cities, so that the other public sectors and cities can learn from them; what strategic planning issues are relevant to enhance e-participation under the smart cities framework?; what strategic decisions must be taken to promote and improve the initiative of Smart Cities?; what incentives can enhance the co-creation initiatives in Smart Cities?; and are there any risks in this kind of initiatives?

Furthermore, because of the certain maturity acquired by e-Government field in term of research, the most recent studies have tended to be empirical and this has favored collaboration between multidisciplinary fields researchers, resulting in a great number of articles published in international journals with a strong impact factor, as well as a great deal of significant collaboration with public managers and politicians who have contributed their professional

ACKNOWLEDGMENT

This research was carried out with financial support from the Ministry of Science, Innovation and Universities (Spain) (Research project number: SmartGov_Local RTI2018-095344-A-100) and the Centre of Andalusian Studies of the Regional Government of Andalusia (Spain), Department of Presidence, Public Administration and Domestic Affairs under Grant (PRY137/19).

REFERENCES

Aham-Anyanwu, N. M., & Li, H. (2017). E-State: Realist or Utopian? *International Journal of Public Administration in the Digital, 4*(2), 56–76. doi:10.4018/IJPADA.2017040105

Ahn, M. J., & Berardino, M. (2014). Adoption of Web 2.0 by the State Government: The Role of Political Environment and Governors. *International Journal of Public Administration in the Digital, 1*(1), 56–73. doi:10.4018/ijpada.2014010104

Aladalah, M., Cheung, Y., & Lee, V. C. S. (2018). Towards a model for engaging citizens via Gov2 to meet evolving public value. *International Journal of Public Administration in the Digital, 5*(1), 1–17. doi:10.4018/IJPADA.2018010101

Alcaide Muñoz, L., & Garde Sánchez, R. (2015). Implementation of e-Government and Reforms in Public Administrations in Crisis Periods: A Scientometrics Approach. *International Journal of Public Administration in the Digital Age, 2*(1), 1–23. doi:10.4018/ijpada.2015010101

Alcaide Muñoz, L., & Rodríguez Bolívar, M. P. (2015). Understanding e-government research. A perspetive from the information and library science field of knowledge. *Internet Research, 25*(4), 633–673. doi:10.1108/IntR-12-2013-0259

Alcaide Muñoz, L., Rodríguez Bolívar, M. P., Cobo, J. M., & Herrera Viedma, E. (2018). Analyzing the scientific evolution of e-government using a science mapping approach. *Government Information Quarterly, 34*(3), 545–555. doi:10.1016/j.giq.2017.05.002

Alcaide Muñoz, L., Rodríguez Bolívar, M. P., Cobo, M. J., & Herrara Viedma, E. (2017). Analysing the scientific evolution of e-Government using a Science Mapping approach. *Government Information Quarterly, 34*(3), 545–555. doi:10.1016/j.giq.2017.05.002

Alcaide Muñoz, L., Rodríguez Bolívar, M. P., & Garde Sánchez, R. (2014). Estudio cienciométrico de la investigación en transparencia informativa, participación ciudadana y prestación de servicios públicos mediante la implementación del e-Gobierno. *Revista de Contabilidad, 17*(2), 130–142. doi:10.1016/j.rcsar.2014.05.001

Anthopoulos, L., & Fitsillis, P. (2014). Trends in e-Strategic Management: How do Governments Transform their Policies? *International Journal of Public Administration in the Digital, 1*(1), 15–38. doi:10.4018/ijpada.2014010102

Atkins, S. E. (1988). Subject Trends in Library and Information Science Research 1975-1984. *Library Trends, 36*(4), 633–658.

Deakin, M., & Reid, A. (2017). The embedded intelligence of Smart Cities: Urban life, citizenship, and community. *International Journal of Public Administration in the Digital Age, 4*(4), 62–74. doi:10.4018/IJPADA.2017100105

Ferro, E., & Osella, M. (2017). Smart City Governance for Sustainable Public Value Generation. *International Journal of Public Administration in the Digital, 4*(4), 20–33. doi:10.4018/IJPADA.2017100102

Forrester, J. P., & Watson, S. S. (1994). An Assessment of Public Administration Journals: The Perspective of Editors and Editorial Boards Members. *Public Administration Review, 54*(5), 474–482. doi:10.2307/976433

Fraser, D. (2000). *QSR Nvivo. NUDIST Vivo. Reference Guide*. Melbourne, Australia: Malaysia. QSR International Pty. Ltda.

Garfield, E. (1972). Citation analysis as a tool in journal evaluation. *Science, 178*(60), 471–479. doi:10.1126cience.178.4060.471 PMID:5079701

Heeks, R., & Bailur, S. (2007). Analyzing E-government Research: Perspectives, Philosophies, Theories, Methods, and Practice. *Government Information Quarterly, 24*(1), 243–265. doi:10.1016/j.giq.2006.06.005

Karamagioli, E., Staiou, E. R., & Gouscos, D. (2014). Government spending transparency on the Internet: An assessment of Greek bottom-up initiatives over the Diaveia Project. *International Journal of Public Administration in the Digital, 1*(1), 39–55. doi:10.4018/ijpada.2014010103

Krippendorff, K. (1980). *Content analysis: An introduction to its methodology*. Sage Publications, Inc.

Lan, Z., & Anders, K. K. (2000). A Paradigmatic View of Contemporary Public Administration Research: An Empirical Test. *Administration & Society, 32*(2), 138–165. doi:10.1177/00953990022019380

Legge, J. S. Jr, & Devore, J. (1987). Measuring Productivity in U.S. Public Administration and Public Affairs Programs 1981-1985. *Administration & Society*, *19*(2), 147–156. doi:10.1177/009539978701900201

Löfstedt, U. (2005). E-Government – Assessment of current research and some proposals for future direction. *International Journal of Public Information Systems*, *1*(1), 39–52.

Maxwell, S. P., & Carboni, J. L. (2017). Civic engagement through Social Media: Strategic stakeholder management by high-asset foundations. *International Journal of Public Administration in the Digital Age*, *4*(1), 35–48. doi:10.4018/IJPADA.2017010103

Navarro Galera, A., Rodríguez Bolívar, M. P., Alcaide Muñoz, L., & López Subires, M. D. (2016). Measuring the financial sustainability and its influential factors in local governments. *Applied Economics*, *48*(41), 3961–3975. doi:10.1080/00036846.2016.1148260

Nord, J. H., & Nord, G. D. (1995). MIS research: Journal status and analysis. *Information & Management*, *29*(1), 29–42. doi:10.1016/0378-7206(95)00010-T

Norris, D. F., & Reddick, C. G. (2013). Local e-government in the United States: Transformation or incremental change? *Public Administration Review*, *73*(1), 165–175. doi:10.1111/j.1540-6210.2012.02647.x

Ohemeng, F. L. K., & Ofosu-Adarkwa, K. (2014). Promoting transparency and strengthening public trust in government through Information Communication Technologies?: A Study of Ghana's E-Governance Initiative. *International Journal of Public Administration in the Digital*, *1*(2), 25–42. doi:10.4018/ijpada.2014040102

Plümber, T., & Radaelli, C. M. (2004). Publish or perish? Publications and citations of Italian political scientists in international political science journals, 1990-2002. *Journal of European Public Policy*, *11*(6), 1112–1127. doi:10.1080/1350176042000298138

Rodríguez Bolívar, M., Alcaide Muñoz, L., & López Hernández, A. M. (2015). Research and Experiences in Implementing e-Government endeavors in Emerging Countries: A Literature Review. In K. J. Bwalya & S. Mutula (Eds.), *Digital solutions for contemporary democracy and government* (pp. 328–346). doi:10.4018/978-1-4666-8430-0.ch017

Rodríguez Bolívar, M. P. (2015). The influence of political factors in policymakers' perceptions on the implementation of Web 2.0 technologies for citizen participation and knowledge sharing in public sector delivery. *Information Polity*, *20*(3), 199–210. doi:10.3233/IP-150365

Rodríguez Bolívar, M. P., & Alcaide Muñoz, L. (2018). Political idelology and municipal size as incentives for the implementation and governance models of Web 2.0 in providing public services. *International Journal of Public Administration in the Digital*, *5*(1), 36–62. doi:10.4018/IJPADA.2018010103

Scholl, H. J. (2009). Profiling the EG research community and its core. *Lecture Notes in Computer Science*, *5693*, 1–12. doi:10.1007/978-3-642-03516-6_1

Webster, J., & Watson, R. T. (2002). Analyzing the past to prepare for the future: Writing a Literature Review. *Management Information Systems Quarterly*, *26*(2), 13–23.

Wimmer, M. A., & Sherer, S. (2018). Supporting Communities through Social Government in Co-Creation and Co-Production of Public Services: The SocialGov Concept and Platform Architecture. *International Journal of Public Administration in the Digital*, *5*(1), 18–35. doi:10.4018/IJPADA.2018010102

Yildiz, M. (2007). E-government research: Reviewing the literature, limitations, and ways forward. *Government Information Quarterly*, *24*(3), 646–665. doi:10.1016/j.giq.2007.01.002

Chapter 2
Human Factor and ICT Use in the Context of Modern Governance

Uroš Pinterič
Alexander Dubček University in Trenčín, Slovakia

ABSTRACT

The chapter presents the development of the e-government in the case of Slovenia, taking in the consideration the human factor as main obstacle. On the side of the citizens as well as on the level of public administration, there is misconception of the purposes of the ICT, and thus, it appears that both partners in this context communicate past one another. In this manner, it exposes the question of the motivation, which is further supplemented by the survey results from Slovenia, showing general lack of motivation measured through the ignorance of the technology potentials as well as of existing threats. The main argument of the chapter is that lack of motivation will block any reform attempt by creating negative human environment, as well as wrong motivation to use ICT in administrative communication will result in sub-optimal or abusive use of the technological potential.

INTRODUCTION

Reforming public administration towards classical 3E model (increasing effectiveness, efficiency and economy of their services) is a long term goal of administrative science in combination with other scientific fields. What seems to be managerial question at the first glance, has in fact strongly ideological roots of modern state, where economic consumerism is replacing the rule of the law principle. New

DOI: 10.4018/978-1-7998-1526-6.ch002

public management principle is replacing classical Weberian bureaucracy and then shifts towards neo-Weberian state approach again. While classic bureaucracy was strongly resting on legal legitimacy and rule of the law (sometimes to the point "ad absurdum"), new public management (NPM) demanded respect for the law but with understanding that different personal interests have right to exist and to be followed within the legal regulation as well as more effective, efficient and economical way to deliver public services should be developed. This often led to the privatization of public goods where neo-Weberianism in mid 2000s started to defend re-introduction of the state (Drechsler, 2005; Drechsler and Kattel, 2008; Pollitt and Bouckaert, 2011) and recognised that after all state cannot keep just minimal role of taxing the population for actually providing essentially nothing (since NPM privatize practically everything from healthcare, education, research, development, security and in most absurd cases partially even military), but existence of the political system. However, some authors (e.g. Dunn and Miller, 2007) argue that the Neo-Weberian state is more of a criticism of the previous state management approaches than an answer to the everlasting issue of balance between public goods and their costs. Kuhlmann at al. (2008), on the other hand, offer arguments that, even in the established democracies with long bureaucratic traditions, administration cannot change its practices so easily and demands the strong role of the state back. On this ground idea of good governance (e.g. Klimovský, 2010) and later open government (e.g. Grimmelikhuijsen, Feeney, 2016), which shall include computer-mediated transparency (Meijer 2009), website information provision (Grimmelikhuijsen, Welch 2012), financial transparency (Pina, Torres, Royo 2010), and online participation (Feeney, Welch 2012; Ma 2014; Oliveira, Welch 2013). Based on this one can gain the understanding of government accessibility, transparency, and participation. However, it is very hard to overlook that good governance as much as open government concepts are just re-packing the basic principles of efficiency, economy and effectiveness combined with the desire for transparency (when allowed by "higher interests") and participation (when not interfering with "higher interests" of political elite).

Under such circumstances of transition from strong to lean administration and back, the technological development introduced the concept of the e-government/e-governance and different correlating terms from smart cities (which under different understanding exist already before) to e-democracy and participative citizenship. States, which believe that they are following the trends (if not even setting them) are in recent two decades jumping from one temporarily idea to another, creating more confusion among the population than anything else. However, they are at the same time often refusing ideas coming from citizens and show great fear from effects of their own concepts.

Chapter tries to show on the case of Slovenia, two-faced reality of the limited change potential in the practice. In the perspective of administrative reforms,

Slovenia is one of the countries following all modern trends from NPM to good governance, different e-supported concepts, including smart cities. All this in order to achieve 3E and to satisfy the need to be recognised as modern, democratic and customer oriented administration. However, the results are limited at best. Two main reasons are lack of political motivation for a change (which is necessary for change to actually happen) and lack of proper and honest evaluation of the certain issue prior and after introducing the changes (which is necessary to know what and how to change). Common denominator of both issues is lack of motivation for the actual change. Despite it is rather simple to accuse (even if justified) the politics and public administration for all bad that happens, there is deeper issue behind the ICT use in the society. Politicians are elected among citizens as well as civil servants are just citizens working for the state, which means that they are equals to the general population (with the potential to be hidden behind the power of the sovereignty of the state in relation to their peers). However, this means that one needs to question the general motivation for the change and general understanding of the concepts. It can be assumed that citizens have no motivation and no knowledge of the modern administrative changes, which makes them ignorant towards the potential that certain changes could bring. Consequently top-down introduction of such changes (without public deliberation and involvement) would be nothing more but waste of budgetary (citizens') money.

In this perspective, ICT environment of administrative changes was tested by the questionnaire in second largest Slovenian city (Maribor) in order to understand potential success of introducing ICT driven administrative concepts, such as smart cities. Selection of Maribor is based on its non-capital level combined with the size which can be still considered comparable with medium cities of Europe (urban environment in which use of ICT in administrative context has the most effect). The results were put in the context of the administrative / state side of the e-government reality in Slovenia.

MOTIVATION FOR A CHANGE

As it was indicated earlier, one of the crucial problems of any administrative change is in fact motivation. By stating this, we assume that any change is possible, if there is interest/motive to do it. In the opposite case, change will not be introduced or it will be jeopardized to the level of absolute inefficiency. In the perspective of political change management, it is possible to differ two levels of the interest for a change (superficial and real interest/motivation). Superficial motivation for change can be most accurate described as "empty words" which only create budgetary expenses with (sometimes even expected and desired) no effect. The real motivation/interest for change is the one where responsible actors expect some changes in certain field or policy.

In order to understand motivation we are returning to the classical concept which has explanatory power to help us understand why so many political changes are about to fail. This can be supported also by IT specialist dealing with smart cities, Robinson (2013), who agrees that the Maslow (1954) hierarchy of needs can be used as an appropriate base. According to Maslow (1954), all our needs are shaped in a pyramid structure, based on the number of people who have certain »need«, and where each next need is characteristic of fewer people. According to this hierarchy, all people have biological/physiological needs, such as for food, air, etc., Rather universal are also needs for security of body and individuality, including social security. The third level of needs is the need for belonging, socialization, love, which is not as much universal as previous sets. The needs for recognition and esteem represent the next level, and the self-actualization needs form the last level, which is common only to a smaller share of members of any society. Regardless of which of these needs are characteristic of any individual, it can be argued that more of the needs from an individual's list are fulfilled, better individuals assess the quality of their lives.

As it can be understood, some aforementioned needs are developed and fulfilled by individuals. Other needs are arising out of social interactions and can be fulfilled in interaction with others (individuals or institutions). In this perspective, smart cities can be seen as a possibility for individuals, to report their needs and expect the local authorities to help people to satisfy them, as it was observed in the case of Singapore (Mahizhnan, 1999). This approach goes along with the idea of participatory governance as part of today's mainstream politics (e.g. Linders, 2012). However, as it was noted before, public sector has limited resources and it is expected that solutions will demand also the participation of the business and associational sectors (Lovan et al., 2005b). Public participation, no matter how strongly motivated, is not a cure-all tool. Recently it was pointed out that there were some shortcomings which made public participations' outcomes very questionable (e.g. Mosse (2001), Cleaver (2001) or Beall (2005). Under the neo-Weberian wave, the state has expanded its activities into too many fields, but did not improve efficiency and often wasted the resources. Governments' failure is as well frequently linked to the fact that sub-optimal results serve the interests of certain politicians and government officials (Mitlin and Satterthwaite, 2004; Coursey and Norris, 2008; Paulin, 2013). New allocation of competences between government and society is needed in order to give citizens more responsibilities and possibilities to act on their own. There is a need for more opportunities where initiatives of citizens can be developed (Schultz, 2001). Although it seems as a great combination, allowing more participation and less state influence, in practice it is often just the other way around. New technologies, empowering citizens, are in fact enabling the authorities to manipulate citizens according their ideas of ruling the territories (e.g. Pan et al., 2013).

The ICT supported administration would mainly influence two different types of human needs according to the Maslow hierarchy (1954): security and self-actualization. The security aspect is strongly connected to the surveillance and in this case technologies can raise the feeling of security only if the predominantly defined undesirable behaviour is reduced – crime (thefts, murders, etc.) as well as delinquency (public urinating, alcohol abuse in public spaces, etc.). Quality of life in the sense of higher security will increase as long as citizens will feel more secure than controlled. This will happen much faster when citizens do not notice any change in their personal security (Leman-Langlois, 2008).

According to the Maslow's hierarchy of needs (1954), self-actualization in smart cities can be seen as empowerment of citizens who are willing to participate. Authorities provide inhabitants with opportunities to improve their living habitat by suggesting various activities, actions or changes that should be carried out by the authorities or the community itself (Kim and Lee, 2012; Linders, 2012). For participating individuals, quality of life could increase when their suggestions are not only taken into consideration, but they are also accepted and implemented. If their ideas are not even discussed or if all suggestions are rejected, then the individuals will understand such behaviour as a loss of time (Mahrer and Krimmer, 2005; Islam, 2008).

In order to reach such environment, open for changes and participation another motivation for a change is in question. Administrative science refers to it as administrative culture and is the informal backbone of the administrative practice, which can create change accepting or change rejecting environment). Rman and Lunder (2003: 108) state that administrative culture can be often one of most relevant factors of successful work in public administration. Saxena (1996: 706) defines administrative culture as pattern of values and expectations that are common to all members of some organisation. Expectations and values create rules (norms) that very effectively create appropriate behaviour of individuals and groups in organisation. Older than administrative culture is, more values and norms are rooted and changes are harder to be carried out. At the same time this is also the greatest barrier to changes in organisation. Saxena (1996: 705) presents special model with all elements that should be reformed in order to reform public administration.

Saxena (1996: 706) argues that bureaucratic rigidity, hierarchy and in some cases even autocracy are main reasons for bad solutions. As example case when civil servants are strongly supporting value of paper documentation and archiving and who find use of electronic document too abstract for use in practice is exposed. Saxena (1996: 706) argues that, despite technological innovations, changes are not easily and quickly introduced. The main reason for such situation is existing administrative culture that needs change in order to change of strategy of acting in order to introduce new technologies and finally also to adjust administrative structures as it was noted.

Klimovský (2008: 182-184) shows, how within the formal hierarchical structure of organization is over driven by informal patterns of interpersonal communication that can disturb organizationally predefined communication flows. These patterns can form specific informal structures, which are able to block institutional routines.

Different authors support idea that introduction of the ICT in the administrative processes will speed the reforms in the other spheres of public administration. However, West (2004: 24) warns that science should not accept interpretation, that e-government easily delivers changes in public services, democratic responsiveness and citizens' trust in public administration, without serious research. Among evidences against such technology initiated changes of administrative culture and administrative processes West (2004: 24-25) shows USA example of non-interconnect and non-integrated web pages of public administration and lack of standardization of navigation tools. West (2004: 25) compares this situation to Babylon tower that completely failed as a project, because of too great need for megalomania and incompatibility of languages. 6 (2004: 57) argues that even civil servants themselves often admit that one of the greatest problems at use of ICT is lack of knowledge in this area, lack of readiness to learn new things and inappropriate administrative culture that should support complete use of tools offered by the new technologies. Also Rman and Lunder (2003: 110) are paying more credit for changing public administration to human resource management, and not to the other factors such as introduction of ICT and changes of working processes connected to ICT. According to mentioned, we can believe that change of administrative culture is precondition for effective reformation of public administration. However, inappropriate use of ICT can persuade civil servants in old bureaucratic patterns of work, by not changing the nature of the work processes. Slovenia experienced such case after introduction of e-application for personal income tax, which was announced as big hope and after few years ended up as publicly badly accepted project. On its peak of "success" under 20.000 out of about 1.700.000 taxpayers used e-PIT. The reason was two-folded. On one side, there was still relatively low penetration of proper technology combined with all possible issues of digital divide. On another hand the government did not provide any incentives to those who would submit the PIT report electronically (Pinterič, 2015). Their reports were considered absolutely just as paper version, despite electronic from would allow immediate calculation of the tax (not even to think about smaller bonus or anything similar).

However, we are not able to ignore the general fact that modern generations are more and more used to the use of the ICT for different purposes, regardless of the "side-effects" and so is the state (for obviously different reasons). Internet of

Things as technological backbone of its soft product, smart community, penetrates our daily life more and more and enables more and more not only the technological benefits to individuals but also changes the identification of them in more general sense (e.g. Kim, 2017). Smart community is thus not only ICT enriched community but it seems to be mainly the response to initial alienation of the technology society, which became not only individualized but also isolated. Smart community thus return the sense of belonging (e.g. Li et all, 2011) and recreates certain form of "imagined community" (Anderson, 2006). Despite the general attention of the chapter is the administrative procedures related ICT use, we cannot skip the general perspective on ICT functionality. General motivation can be seen in simplification of the life and life tasks (in some cases, such as for the handicapped people, justified). Ideology of "on-line" life (which shall not be confused with the virtual life), multiplies and yet simplifies our networking capacities, it enables multitasking and multi-careerism, even in the circumstances of physical immobility. It gives false impression of the omnipresence and self-centrism. However, it only fertilises the space for bipolar disorder, anxiety, narcissism, multiple personalities disorder and entitlement. Not for everyone, but for many people. Lack of the knowledge and awareness of the actual functionality of "on-line" environment, makes people more vulnerable and public while they believe they are able to be anonymous and protected. (on different negative psychological effects of ICT (Oliver, Rayen & Bryant, 2020).

On the other hand, state is keen on using the ICT officially for improving the services for the citizens, while actually trying to reduce the costs. State is announcing new and more flexible services, while in a first stance (due to the historical necessity or by the fact that political power corrupts by itself) increases the surveillance over the population and controls its behaviour by adjusting the regulation in order to maintain the political stability.

Despite this paragraph of critical reflection should be taken with some reservation and especially on the case to case basis, the tendency is rather clear, when we dare to compare the intentions with the effects.

SLOVENIAN CASE IN PRACTICE

In order to understand how much can citizens add up to the improvement of the quality of life by participation, questionnaire with close-ended questions, was randomly distributed among 100 citizens on the streets of Maribor. Maribor is non-capital city of Slovenia, which can be considered European comparable urban environment which is open to and can be positively affected by ICT supported administrative services, if in interest of inhabitants. More than 50 per cent of the interviewed were from the city, the rest came form surrounding areas. Concerning the sex distribution, one can speak of an approximately representative sample (51.7 per cent female respondents, which is similar to the national sex distribution). The age structure was normally distributed, with a majority of the respondents between 31-60 years of age (69.1 per cent). Distribution of education shows similar shares had secondary and tertiary education (37-38 per cent each). The respondents were office workers (18.8 per cent), the unemployed (18.1 per cent), production workers (16.8 per cent), pensioners (16.1 per cent), and civil servants (12.1 per cent). Other categories of "occupation" in the individual countries are represented by shares smaller than 10 per cent. According to the survey, in Maribor there are 59.7 per cent of daily Internet users. Despite statistical extrapolation of the conclusions on national level cannot be done, the sample shows enough comparability that we dare to assume that similar answers would be gained also in the national context.

Empirical Results: Public Perception of Technology, Trust and Control

In the first part of our research we wanted to find out how citizens use technology. Long list of the strategic documents on different levels; from the UN, EU to the national level (e.g. European Commission 2010 and European Commission 2012) demand (and somehow recognise Slovenia as) information society, there is a significant question of how information and communication technologies are used in reality. In Maribor, the use of the Internet and e-mail is strongly and statistically significantly connected with age, education and work (younger, more educated and office workers will be using these technologies more often). At the same time, e-government and e-banking are strongly and significantly connected only with work (in all cases, the Pearson correlation coefficient is between 0.35 and 0.6 with the correlation significant at the 0.01 level). At the same time, such significant correlations are achieved in the multi-tasking use of mobile phones only in the case of age younger generation uses mobile phone for more different activities). This can indicate also that a digital divide (predominantly based on age) is not completely overcame in Slovenia.

As it was mentioned, the Internet is daily used in almost 60 per cent of the cases. The use of the e-mail is also regular (54.3 per cent). At the same time a majority of the respondents (40.3 per cent) never use e-banking as well as the e-government. Even more disturbing is that many respondents argue that they do not know what e-government is.

Concerning the use of the mobile phone, almost all respondents use it for calling and texting. When it comes to the use of the mobile phone for other activities such as net browsing, e-mail use, mobile banking or administrative purposes, the share is lower than 30 per cent. However, over 30 per cent of the respondents use the mobile phone for fun. Based on this, it can be argued that citizens are not interested in using modern technologies for more demanding tasks.

At the same time, trust in technologies is an important factor for any ICT driven politico-administrative concept to be successfully introduced. Despite the trust is ungrounded, since it is evident that legal and legitimate postulates of privacy, anonymity and ethical behaviour are systematically violated by the states and service providers, it is essential for "e-concepts" to be implemented. In order to understand the general concern about privacy, respondents were asked how they felt about sending private data to authorities, then how they sent privacy sensitive data and to whom they would send such data using information and communication technologies.

On one hand, 60.8 per cent of the respondents want to know how their personal data will be used by local authorities if they are requested (Table 1).

Table 1. Relation towards management of personal data by local authorities

	Slovenia
want to know	60.8%
care	27.7%
ignore	10.8

Source: own research

But on the other hand they act objectively irresponsible by having no problem to send private information via the mobile phone or internet (60.7 per cent already did so). The respondents in Maribor often send privacy sensitive data to their friends (what shows a high level of interpersonal trust, but they can be in some cases considered irresponsible) but not so often to different institutions. Surprisingly, relatively many respondents send personal data also to those people who only claim that they have the right to know (25 per cent) (Table 2).

Table 2. Sending personal data to someone who only claim they have the right to know them

	Slovenia
no	75%
yes	25%

Source: own research

Despite such behaviour is irresponsible it shows high level of trust that can be abused or used for implementation of more ICT supported administrative processes.

On the other side of trust is control which is partially needed for the functioning of the services but often extended to the abusive invasion of privacy of the individuals. The respondents were asked about their opinion on the following statements (and thus tested in their knowledge of the ICT control potential): all information activities can be tracked; a computer can be monitored when it is connected to the Internet; and a computer from which e-mails are sent can be tracked (when, although illegally, all can be true). Then they needed to say whether their mobile phone can be located when it is switched on, switched off, and when the battery is removed (in this case it is technically impossible to track the mobile phone only when the battery is completely removed).

Results of the survey show no significant differences in recognising different security and privacy risks in relation to sex, age or education, but some weak tendencies can be seen in the case of the older generation more often recognises a switched-off mobile phone as still traceable. However, these differences can hardly be connected to age as an independent factor (Cramer's V or Phi is less than 0.3).

For additional comparison, we took the information on the Internet use frequency and correlated it to different types of control. From the collected date it is not possible to indicate any statistically significant correlation. In this manner, one can argue that the use of technology increases the level of knowledge or at least the awareness of the privacy risks.

The relation towards control was measured by interconnected questions, and the respondents were asked to clarify who was responsible for providing data to local authorities, if local authorities had the right to control population, and if authorities needed to do so. Based on survey, it is possible to argue that people who agree that local authorities have the right to control all citizens will very much likely agree also with "the fact" that that authorities need to do so. At the same time, many of these people also agree that citizens should inform authorities of changes of all relevant data on their own accord. On the other hand, people who think that authorities have no right to control citizens will also most likely see no need for control and in many cases they will oppose the idea that local authorities should be informed of different changes of citizens´ personal issues (it is assumed that such changes are important for authorities in the first place, e.g. a change of address).

Most of the respondents believe that data should be provided to authorities on request (59.1 per cent) (Table 3).

49.7 per cent of the respondents believe that municipalities have the right to control only suspicious (not defined what they are) activities. Same share (49.7 per cent) of the respondents believe also that there is need to control suspicious activities. Most of the other respondents oppose any right or need for control (see Table 4, Table 5).

Table 3. Need for reporting personal data changes to authorities

	Slovenia
always provide data	13.4%
provide data on request	59.1%
not needed	27.5%

Source: own research

Table 4. Authorities' right to control

	Slovenia
right to total control	14.8%
right to control suspicious activities	49.7%
no right to control	35.6%

Source: own research

There is no statically significant relation between the sex, age, current occupation or education and the relation towards control in any of the analysed countries. In both cases, both men and women responded equally as for control of local authorities. In this sense it is not possible to argue that there is concern about the control. Most of the respondents believe that a certain level of control is appropriate especially if something is marked as suspicious behaviour (even when it is not defined, which leave broad space for prejudices and personal interpretations).

The survey results mainly show general indifference towards the technology use and potential as well as towards privacy and control. Respondents feel satisfied with a certain level of control without any special questioning what this control means for them as for citizens. At the same time, they show strong ignorance about their personal data management, which gives us somehow the feeling that the authorities can use technologies in any way they want as long as they do not limit citizens in their daily behaviour or as long as they do not request any particular activity from

Table 5. Authorities' need to control

	Slovenia
need to control everything	13.4%
need to control suspicious activities	49.7%
control forbidden	36.9%

Source: own research

them. Such stance towards the question of control opens new perspective on the value system embedded in the modern society, where the question of privacy is pushed aside, usually in order to achieve better level of security. However, the critical approach to the life shows that increased control over the time did not improve the level of safety for the general public in major security situations (poverty, criminal, road safety). On the other hand it increased the risk of privacy invasion, identity thefts, industrial and state spying for different reasons.

On the other hand, the early research on Slovenian administration response to the citizens communication initiation it was found out that already between 2003 and 2007 the response was strongly correlated with the type of the institution. While local government institutions (municipalities) were the weakest link, the administrative units (which were and still are predominant address for administrative procedures in the case of Slovenia) were the most responsive (over 90%). However, Slovenian administrative system never managed to unify their services for those, who are accessing them via ICT. Slovenia has even nowadays rather peculiar system of enabling people to participate with more easy with the political and administrative system (see Pinterič, 2015). Despite the concept of one-stop-shop was developed few decades ago, ICT enables merging of the services, and even if the mid-2000s system of e-government portal had predominantly two sub-systems (one based on mid-1990s structure, and other partially user-friendly, based on life events, merging the events, legal information and proper forms), it was transformed in three not properly interconnected portals of life situations, legal backgrounds and proper forms. At the same time, in the situations when state has the possibilities to (due to the possible connectivity of the databases) act in speedy manner, they do very little effort to reduce the legal deadline of 30 days. In the case of complex situations, rather simple situations take potentially more than a year (e.g. proper calculation of personal income tax for a Slovenian citizen who is simultaneously and independently employed in Slovenian and foreign institution). In the meantime, Slovenian state introduced necessary e-tax reporting for all legal subjects (regardless of their size, based on number of employees, profit or revenues). Also so called tax cashiers (directly connected to national tax office, where the transaction is registered for later inspection if proper taxes were paid) were introduced for all cash transactions, resulting in the absurdity that school market transactions were subject of inspection and tax evasion charges (this was later dismissed and legally covered by the exemption of volunteer organisations, which can reach yearly revenue less then 5000€). Fact that such situation occurred in the first place, shows that ICT tools are introduced rather selectively, based on the needs of administration, while the citizens are still trapped between paper forms and personal visits to the proper office within the official hours.

DISCUSSION

Despite this short case should be concentrated on empirical data and analysis of civic and bureaucratic side in administrative communication environment and their background rationales for their behaviour, it is mainly about the new digital divide and potentially new societal order. If we can argue that the classical digital divides are disappearing in the developed world, we can potentially observe the formation of the new one. It is not technology accessibility related, it is not age, knowledge or use related. In the times of political correctness, fake news and weakness as major concepts shaping the modern developed societies, it is identity related. It is the question of who am I, and what is my stance towards the ICT. This principle might be seen as logical development of awareness based use of the ICT. But it seems the final break between real and virtual society, which goes as far as changing not only the means but the patterns of the communication. It establishes new professions (while still maintaining old industries), such as influencers (which cannot be understood as elaborated version of marketing). And it creates simplified environment, which is based on the assumption that nothing ever fails. Smart world (internet of things, etc.) works on the false assumption that technology never fails, that the world is absolutely secured. However, the real world practice showed us that even most sophisticated security systems are penetrable (regardless if state or private), that identity theft is not only possible but likely and that it can be done convincingly enough to fool the biometric security systems, and that more sophisticated technologies are more expensive to repair and more likely to require expert maintenance. The new digital divide can be seen as forming around the ignorance of aforementioned issues, arising from the technological imperfection.

Second issue was partially raised already before. Different expectations of different subjects what is the role of the ICT in the modern society. Despite we can all agree on smart technologies (even when recognising their flaws) as the modern reality, we have rather different perspectives, what is the definition of the "smartness" in the technologies and what this means for the users. Smart technologies can be seen as technologies which are able to improve human life due to their learned characteristics. First issue with the smart technologies is that under some very basic circumstances (e.g. power shortage) they lose their functionality. Second, they are not prone to malfunction and in many cases have weak protection in the combination with the vulnerability of the data which is ensuring their functionality. Third, self-protecting measures, taken by the alert individuals, might backfire in certain legal systems (e.g. dash-cams use in Austria is practically illegal, even if it would potentially serve an individual to protect oneself). On the other hand European countries in general are free to use and exchange data on vehicles registered in the traffic rules violations, without the consent of the individuals. This argument can be easily dismissed by the question of public and private interest, however, the United Nations Declaration of human rights in 12th article clearly protects one's privacy which might be well understood as protection form the state surveillance as well. In order to excuse the erosion of the one's privacy, due to the predominantly state measures (since they are organized, systematic and might have broader consequences than individual voyeurism), states promote such ICT supported measures by need for increased security. However, the later never comes. Social and economic security is decreasing in many countries, the general safety varies and different statistics shows, that despite there are better methods of recognition the share of solved crimes is decreasing. Under such circumstances, all the biometrical information collection might have different purpose than solving the crime. The argument behind is, that the repressive apparatus of the state is becoming the protector of the state from the citizens by surveillance and subsequent repressive measures if needed, while protection of the citizens is often the secondary tasks, predominantly done in order to avoid the civic unrest.

Each individual (as a private person) still retain the right to decide how much more than legally required amount information about oneself will feed into different databases. And for which price. While business subjects usually still have to provide us with something in exchange for access to our life, state can demand it by law or simply takes it in the perspective of own sovereignty. In this sense smart technologies can get completely new perspective and indirectly imply direct or indirect stupidity of the individuals (being as reduced level of knowledge, such as ability to read maps, hand-write and other skills which can be proven useful in the situation of the technological failure, or as falling for more subtle traps, reducing our freedoms beyond the inevitable level). Part of scientific community (for overview see Bunz, 2015) is addressing the question of positive and negative effects of simplification of the ICT use and life behind on our mental capacities. It seems that general agreement is reached that introduction of ICT decreases our functional smartness and potentially leads to different psychological issues.

CONCLUSION

Bases on the theoretical debate and short empirical research one can draw following lines of the subject discusses in the text. Any change on any given level is done only by the support of certain interest and is thus motivated. However, most of the administrative changes connected to the reform of public administration from more bureaucratic one towards more effective, efficient and economic one lacks any proper motivation on the level of the individuals who are part of it since, it has no clear positive effects in a sense of life security. In fact it promises economic threat to unknown number of employees who stand in opposition to very few of those who would reach better self-actualization by conducting such reform. At the same time any suggestion on changing of bureaucratic procedures has no immediate positive effect on citizens (no significant immediate tax reduction was ever promised, in some cases not even faster procedures), which means that change is considered only as unnecessary modification of the process which does not relate to them. Short empirical survey in second biggest city in Slovenia shows also few other elements blocking the reforms of bureaucratic nature of Slovenian public administration. Assuming that the most effective changes today are related to the use of the ICT, we can argue that there is still significant group of people more than decade after first Slovenian e-government strategy. We can even introduce new digital divide (which should be systematically tested), between those who use basic functions of ICT and those who use full extend of functions which are available. In this context, we can argue that majority of population uses just simple functions of certain technology (internet: browsing, email; mobile phone: calling, texting) while more advanced functions are reserved for minority of population (decision-makers and other types of self-promotors). Based on this limited use of technologies by majority of population it is no surprise that the survey showed also weak knowledge on threats to the individuals as well as low awareness of empowerment potential (measured as demand for higher privacy, government responsibility and lower control).

All in all, it is possible to argue that society misses the motivation to demand changes towards ICT driven government with realized participatory potential. This can be connected to the fact that society which is using information technologies is in fact not automatically information society, and thus our expectations are too high. Or that citizens (among them also civil servants) are subconsciously aware that significant reform of bureaucratic model of administration towards e-administration will not only reduce the employment of civil servants for at least one third (eliminating all positions connected to logical control and approval), but also reduce the possibility for different exceptions from the rule, based on discretion of the civil servant.

New concepts and paradigms (e.g. open government, smart cities, smart communities), which are emerging based on the development of the society are raising many questions. Some of them are derivates from the perspective how to make it work, but much more importantly; what is new. Societies as a whole as well as their individual parts never in history faced so many rapid changes, requesting equally rapid responses that value of new paradigms and concepts looses its explanatory value since it is often nothing more as new angle of addressing very same issues (and often neglecting the possible new solutions due to the timeless laws of human behaviour in the sense of satisfying own basic needs.

On the level of the presented empirical research, further research should be directed towards national survey, despite we can assume that general outline of the results will remain rather similar, and can differ between rural and urban territories and can additionally differ based on education and age, which was not further detailed in this chapter. In this perspective, further research should be concentrate in the unconventional fields, questioning the real needs of the citizens in relation to technology as well as to the state, since both IT companies as well as the state often create services and only afterwards they push demand by closing similar services or by making certain service compulsory. At the same time, discussion on the motivation for the use of ICT services should be also discussion on transformation of the society, not as much into information society as of technology depending society, creating social detachment and addiction which enables authorities to control behaviour of individuals easier than ever.

REFERENCES

Anderson, B. (2006). *Imagined communities: Reflections on the origin and spread of nationalism*. Verso books.

Beall, J., & Hall, N. (2005). Funding Local Governance: Small Grants for Democracy and Development. Bourton-on-Dunsmore: ITDG Publishing. doi:10.3362/9781780443287

Bunz, M. (2015). School Will Never End: On Infantilization in Digital Environments—Amplifying Empowerment or Propagating Stupidity? In *Postdigital Aesthetics* (pp. 191–202). London: Palgrave Macmillan. doi:10.1057/9781137437204_15

Cleaver, F. (2001). Institutions, Agency and the Limitations of Participatory Approaches to Development. In B. Cooke & U. Kothari (Eds.), *Participation, the New Tyranny?* (pp. 36–55). London: Zed Books.

Coursey, D., & Norris, D. F. (2008). Models of E-Government: Are They Correct? An Empirical Assessment. *Public Administration Review*, *68*(3), 523–536. doi:10.1111/j.1540-6210.2008.00888.x

Drechsler, W. (2005). The Re-Emergence of 'Weberian' Public Administration after the Fall of New Public Management: The Central and Eastern European Perspective. *Halduskultuur*, *6*, 94–108.

Drechsler, W., & Kattel, R. (2008). Towards the Neo-Weberian State? Perhaps, but Certainly Adieu, NPM! *NISPAcee Journal of Public Administration and Policy*, *1*(2), 95–99.

Dunn, W. N., & Miller, D. Y. (2007). A Critique of the New Public Management and the Neo-Weberian State: Advancing a Critical Theory of Administrative Reform. *Public Organization Review, 7*(4), 345–358. doi:10.100711115-007-0042-3

European Commission (2010). *Communication from the Commission: Europe 2020: A strategy for smart, sustainable and inclusive growth.* March 2010 (EC(2010) 2020).

European Commission (2012). *Communication from the Commission: Smart Cities and Communities – European Innovation Partnership.* July 2012 (EC(2012) 4701).

Feeney, M. K., & Welch, E. W. (2012). Electronic Participation Technologies and Perceived Outcomes for Local Government Managers. *Public Management Review, 14*(6), 815–833. doi:10.1080/14719037.2011.642628

Grimmelikhuijsen, S. G., & Feeney, M. K. (2016). Developing and Testing an Integrative Framework for Open Government Adoption in Local Governments. *Public Administration Review, 77*(4), 579–590. doi:10.1111/puar.12689

Grimmelikhuijsen, S. G., & Welch, E. W. (2012). Developing and Testing a Theoretical Framework for Computer-Mediated Transparency of Local Governments. *Public Administration Review, 72*(4), 562–571. doi:10.1111/j.1540-6210.2011.02532.x

Islam S.M. (2008). Towards a sustainable e-participation implementation model. *European Journal of ePractice, 5*, 1–12.

Kim, S., & Lee, J. (2012). E-participation, Transparency and Trust in Local Government. *Public Administration Review, 72*(6), 819–828. doi:10.1111/j.1540-6210.2012.02593.x

Kim, T., Lim, J., Son, H., Shin, B., Lee, D., & Hyun, S. J. (2017). A Multi-Dimensional Smart Community Discovery Scheme for IoT-Enriched Smart Homes. *ACM Transactions on Internet Technology, 18*(1), 1–20. doi:10.1145/3062178

Klimovský, D. (2010). Genéza koncepcie good governance a jej kritické prehodnotenie v teoretickej perspektíve. *Ekonomicky Casopis, 58*(2), 188–205.

Klimovský, D. (2008). *Základy verejney správy.* Košice: Univerzita Pavla Jozefa Šafárika v Košiciach, Fakulta verejnej správy.

Kuhlmann, S., Bogumil, J., & Grohs, S. (2008). Evaluating Administrative Modernization in German Local Governments: Success or Failure of the 'New Steering Model'? *Public Administration Review, 68*(5), 851–863. doi:10.1111/j.1540-6210.2008.00927.x

Leman-Langlois, S. (2008). The local impact of police videos surveillance on the social construction of security. In S. Leman-Langlois (Ed.), *Technocrime: Technology, crime and social control* (pp. 27–45). Portland: Willan Publishing.

Li, X., Lu, R., Liang, X., Shen, X., Chen, J., & Lin, X. (2011). Smart community: An internet of things application. *IEEE Communications Magazine, 49*(11), 68–75. doi:10.1109/MCOM.2011.6069711

Linders, D. (2012). From e-government to we-government: Defining a typology of citizens coproduction in the age of social media. *Government Information Quarterly, 29*(4), 446–454. doi:10.1016/j.giq.2012.06.003

Lovan, W. R., Murray, M., & Shaffer, R. (2005). Participatory Governance in a Changing World. In W. R. Lovan, M. Murray, & R. Shaffer (Eds.), *Participatory Governance: Planning, Conflict Mediation and Public Decision-Making in Civil Society* (pp. 1–21). Aldershot: Ashgate.

Ma, L. (2014). Diffusion and Assimilation of Government Microblogging: Evidence from Chinese Cities. *Public Management Review, 16*(2), 274–295. doi:10.1080/1 4719037.2012.725763

Mahizhnan, A. (1999). Smart cities: The Singapore case. *Cities (London, England), 16*(1), 13–18. doi:10.1016/S0264-2751(98)00050-X

Mahrer, H., & Krimmer, R. (2005). Towards the enhancement of e-democracy: Identifying the notion of the 'middleman paradox'. *Information Systems Journal, 15*(1), 27–42. doi:10.1111/j.1365-2575.2005.00184.x

Maslow, A. H. (1954). *Motivation and Personality*. New York: Harper & Row.

Meijer, A. J. (2009). Understanding Computer-Mediated Transparency. *International Review of Administrative Sciences, 75*(2), 255–269. doi:10.1177/0020852309104175

Mitlin, D., & Satterthwaite, D. (2004). The Role of Local and Extra-local Organizations. In D. Mitlin & D. Satterthwaite (Eds.), *Empowering Squatter Citizen. Local Government, Civil Society and Urban Poverty Reduction* (pp. 278–306). London: Earthscan.

Mosse, D. (2001). People's Knowledge, Participation and Patronage: Operations and Representations in Rural Development. In B. Cooke & U. Kothari (Eds.), *Participation, the New Tyranny?* (pp. 16–35). London: Zed Books.

Oliveira, G. H. M., & Welch, E. W. (2013). Social Media Use in Local Government: Linkage of Technology, Task, and Organizational Context. *Government Information Quarterly, 30*(4), 397–405. doi:10.1016/j.giq.2013.05.019

Oliver, M. B., Rayen, A. A., & Bryant, J. (Eds). (2020). Media Effects: Advances in Theory and Research. New York: Routledge.

6. P. (2004). *E-governance*. Houndmills, UK: Palgrave Macmillan.

Pan, G., Qi, G., Zhang, W., Li, S., Wu, Z., & Yang, L. T. (2013). Trace Analysis and Mining for Smart Cities: Issues, Methods and Applications. *IEEE Communications Magazine, 51*(6), 120–126. doi:10.1109/MCOM.2013.6525604

Paulin, A. (2013). Towards Self-service Government – A Study of Computability of Legal Eligibilities. *Journal of Universal Computer Science, 19*(12), 1761–1791.

Pina, V., Torres, L., & Royo, S. (2010). Is E-Government Promoting Convergence Towards More Accountable Local Governments? *International Public Management Journal, 13*(4), 350–380. doi:10.1080/10967494.2010.524834

Pinterič, U. (2015): *Spregledane pasti informacijske družbe*. Novo mesto: Fakulteta za organizacijske študije.

Pollitt, C., & Bouckaert, G. (2011). *Public Management Reform: A comparative Analysis: New Public Management, Governance, and the Neo-Weberian State*. Oxford: Oxford University Press.

Rman, M., & Lunder, L. (2003). Organizacijska kultura in javna uprava – priložnost za upravni menedžment. In Konferenca Dobre prakse v slovenski javni upravi (pp. 107-120). Ljubljana: Ministrstvo za notranje zadeve Republike Slovenije.

Robinson, R. (2013). *Can Smarter City technology measure and improve our quality of life?* Available at: https://theurbantechnologist.com/category/smarter-cities/

Saxena, K. B. C. (1996). Re-engineering Public Administration in Developing Countries. *Long Range Planning, 29*(5), 703–711. doi:10.1016/0024-6301(96)00064-7

Schultz, V. (2001). Introduction. In F. Greß & J. Janes (Eds.), *Reforming Governance. Lessons from the United States of America and the Federal Republic of Germany* (pp. 17–18). New York: Palgrave.

Weber, M. (1969). The Three Types of Legitimate Rule (H. Gerth, Trans.). In A Sociological Reader on Complex Organization. New York: Holt, Rinehart & Winston.

West, D. M. (2004). E-Government and the Transformation of Service Delivery and Citizen Attitudes. *Public Administration Review, 64*(1), 15–27. doi:10.1111/j.1540-6210.2004.00343.x

Chapter 3
Diffusion of Innovations Among Mexico:
The Technology Adoption of State Governments

David Valle-Cruz

🆔 https://orcid.org/0000-0002-5204-8095
Universidad Autónoma del Estado de México, Mexico

Rodrigo Sandoval-Almazan

🆔 https://orcid.org/0000-0002-7864-6464
Universidad Autónoma del Estado de México, Mexico

ABSTRACT

The purpose of this chapter is to describe the technological adoption by state governments, based on a longitudinal study of technology in Mexico for which the authors analyzed data from all the local governments from 2010 to 2018. With this data, they proposed a ranking to classify adoption technology, using the diffusion of the innovation theory. They included in the analysis other variables such as the percentage of households with a computer, internet, and other communication technology equipment. The results show that Mexico City is the innovator; Baja California, Sonora, and Nuevo Leon are early adopters, while Oaxaca, Chiapas, and Guerrero are laggards. The most influential variable in the adoption of information technologies is illiteracy, and there is an inverse relationship between technology and illiteracy. Future research will open several paths to understand different adoption behaviors between specific technologies in each state, such as big data, artificial intelligence, internet of things, and smart cities.

DOI: 10.4018/978-1-7998-1526-6.ch003

INTRODUCTION

The adoption of emerging technologies by Mexican state governments is in its earliest stages. Some emerging technologies such as cloud computing, big data, Internet of things, and artificial intelligence are starting to be implemented by governments (Valle-Cruz, 2019). Most of the states in Mexico have a very small advance in the implementation and use of technologies, only making use of static web pages (portals) and social media, even some regions in Mexico do not have Internet services or electricity (SENER, 2017). Despite this, most Mexican state governments are trying to develop and improve portals for service delivery, information dissemination, and implementation of different mechanisms to interact with citizens. The digital divide is a challenge for developing countries (Lu, 2001), because in some regions there a lack of basic technologies like electricity and telephone that avoid the implementation of advanced and emerging technologies. Particularly the Mexican digital divide is a problem of inequality that also reflects the poverty of certain areas in Mexico (Mecinas, 2016).

Regarding social media, it is used by all Mexican state governments to improve interaction with citizens, but the use and adoption of these kinds of technologies have different behaviors for each government (Sandoval-Almazán, Valle-Cruz, & Armas, 2015; Sandoval-Almazán & Valle-Cruz, 2016; Sandoval-Almazán, Valle-Cruz, & Kavanaugh, 2018), because some citizens do not have access to essential technologies and even some people do not even know about them.

However, one of the most important technology uses by state governments to interact with citizens is social media, representing a way to improve government-citizen interaction (G2C); it is a mechanism for dissemination of government activities and information, and it represents an efficient communication channel between government and citizens. Social media is also a tool for citizens to make complaints or petitions to their governments, and it is useful for governments to understand citizens' perception (Valle-Cruz, Sandoval-Almazán, & Gil-García, 2016: p. 1).

In general, there are few empirical studies related to the diffusion of technological innovations in governments (Anderson, Lewis, & Dedehayir, 2015; Chatfield & Reddick, 2018; Wu, J., & Zhang, 2018), and, in a previous research, an explanation was provided to understand, only, the behavior of social media adoptions by governments through the theory of Diffusion of Innovations (Roger, 2003).

Studies related to the diffusion of innovations in government are scarce and this chapter aims to continue with the work done in the article "The Diffusion of Social Media among State Governments in Mexico" published in 2018, where only social media was studied in local governments (Sandoval-Almazán, Valle-Cruz, and Kavanaugh), but analyzing the existing technology data of the Institute of Statistics, Geography and Informatics (INEGI) of the Mexican Government from 2010 to 2018 in order to classify state governments in Mexico based on the Rogers' Theory of Diffusion of Innovations.

The purpose of this paper is to report the technological adoption by Mexican state governments as an starting point for future research in this field. For this reason, the paper focuses on state governments' classification based on the design of a ranking of the Mexican state governments and the diffusion of the innovation theory (Rogers, 2003). This way, we interpreted the technology adoption by Mexican state governments. The contribution of this paper is to classify governments' adoption of technologies in order to design a proper public policy to improve the use of this technology in Mexico.

This paper has been organized into five sections, including this introduction. The second section presents the theoretical framework and review of prior research related to technological factors by state governments and different studies related to the diffusion of innovations. The third section describes the methods we used to collect and analyze technological data from all 32 Mexican state governments. In the fourth section, we present our findings and practical ideas. Finally, in the fifth section, we show conclusions and limitations of the study.

THEORETICAL FRAMEWORK AND PRIOR RESEARCH

This section is divided into three stages. The first section explains the implementation of technologies in government. The second section states the digital divide in developing countries. And the third section shows some works related to technology and the diffusion of the innovation theory.

THE STATE OF TECHNOLOGY IN GOVERNMENT

There are currently a number of technologies that can be implemented in governments, from websites to artificial intelligence. The aim of this section is to expose the state of the art of technology in government at a global level, being aware that in some regions there is a lack of basic technologies such as electricity and telephone.

There are different kinds of technologies that have been adopted by organizations throughout history; one important feature of innovative or emerging technologies is to generate changes in organizations. The positive effects of these technologies have been beneficial in terms of efficiency, transparency, accountability, as well as in the interaction between government and citizens (Valle-Cruz, 2019).

Some of these emerging technologies have become ubiquitous in the public sector. Nowadays, it is difficult to think of a public problem or government service that has not been involved in some substantial way, and the explosion of digital information throughout society offers the possibility of a more efficient, transparent, and effective government (Gil-García, Dawes, and Pardo, 2018). Implementation of technologies needs innovation in governments. For this reason, the public and social innovation sector took a central role in public policies and management debates (Criado, Sandoval-Almazán, and Gil-García, 2013; Karo and Kattel, 2019).

Information technologies are an important component for nations, organizations, governments, and citizens. From a deterministic perspective, good implementation of these technologies reduces costs, improves efficiency, transparency, and generates public value. The scope that information technologies have in governments is important for the delivery of digital services, as well as for interoperability in organizations.

Recently, with a large amount of data produced every day in the big data, and the great variety of needs of the population, an important challenge for governments in terms of technological innovation, is to improve the technological scope to reduce the digital divide and provide better services to citizens. Although in developed countries, information and communication technologies such as television, radio, telephony, and specially Internet are used on a massive scale; in developing countries, they represent challenges in terms of public innovation and the allocation of resources that are useful for improving the reach of population, due to there are some regions without the basic services as electricity and telephone. In this context, technology modify or restructure daily activities and have promising results of improving human conditions.

Specifically, for governments, in Mexico, technological tools based on mobile applications, sensors, online payments, and chats have been implemented to improve government to citizen interaction and government efficiency. The Tax Administration System (SAT) is based on biometric software for the administration of the digital signature (including the support of mobile applications). At the state and municipal levels, mobile applications for citizen attention and service delivery are beginning to be implemented, in addition to different social media mechanisms for interacting with citizens. Also, social media have made their way into government agencies as a channel for citizen communication, and in some municipalities, are emerging as communication mechanisms. Information technologies are essential for the implementation of E-Government, but without the conditions for the implementation of basic technologies, it is impossible to make technological innovations for the benefit of society.

E-Government has been defined as the use of information technology applied in governments (Banerjee and Jain, 2003; Brown and Brudney, 2004; Gil-García, 2012; Luna-Reyes, Hernández and Gil-García, 2009; Palkovits, Woitsch, and Karagiannis, 2003; Reinermann, 2000; Scheider, 2000; Scholl, 2010; Yildiz, 2007). In general, definitions refer to the use of information technology to improve services and information provided to citizens, as well as to increase the efficiency and effectiveness of public management and substantially increase public sector transparency and citizen participation.

The field of E-Government encompasses the use of technology to enable interaction between citizens, governments, businesses, and other organizations. The concept of E-Government was coined in 1995 by the Canadian government in order to ensure the connectivity of the largest number of citizens through information technologies (Arias and Manriquez, 2017). Thus, E-Government, which was created to provide services to citizens, government departments, and employees, involves automating manual documentation processes to translate them into innovative approaches to administration (Carroll, 2005).

Social media offers opportunities for rapid dissemination of information and dialogue with the public, promoting transparency and greater electronic democracy (Magnusson, Bellström, and Thoren, 2012). Organizations have generally incorporated social media elements into their campaigns, product designs aimed for improving the user experience (Tuten, Wetsch, and Munoz, 2015).

A complementary perspective is offered on the role of CIOs and the IT strategy of state governments, that resulted from a case study of the Mérida municipality (Sandoval-Almazán & Gil-García, 2011). Sandoval–Almazán, and colleagues (2012) also analyzed Twitter and Facebook used by all 32 states governments in Mexico, to determine which of them were using Twitter as a tool to communicate with their citizens. Authors created an exploratory collection of Twitter and Facebook data during two months (September and November) in 2010, including tweets, re-tweets, lists and followers, and the number of friends on Facebook -- and provided a model that shows the evolution and use of these social media for politics. In another research with data from 2010-2012, Sandoval-Almazán and Gil-García (2013) used content analysis in the study of two cases – Sinaloa and Yucatán – to determine the emerging use of social media by these state governments. In 2018, Sandoval-Almazán, Valle-Cruz, and Kavanaugh proposed a classification of the 32 Mexico's local governments based on the diffusion of the innovation theory and social media adoption. This paper continues with this research but analyzing information technologies.

Mobile technology is strongly linked to the use of social media. Users invest time in their phones to find information, entertainment, and communication (Au, Lam, and Chan, 2015). For this reason, it is important for governments to have a widespread information technology infrastructure throughout their territory, but in Mexico, some citizens do not know digital government services, and even worse, some citizens do not have the technologies or devices to be connected (Valle-Cruz, 2019).

Similarly, mobile technology has served as a potentiating mechanism for the use of social media by citizens. In this sense, the current trend is about how companies and governments can provide a better social infrastructure through mobile applications and services. The public demand for mobility, as well as the efficiency and productivity from the public sector, is leading to a natural movement from E-Governments to M-Governments (Emmanouilidou and Kreps, 2010; Kushchu, 2007), which translates into the ubiquity of government.

Sundar and Garg (2005) argue that the main feature of these mobile government (M-Government) solutions are "the captured levels required to deliver faster, more cost-effective and scalable services to citizens through mobile technologies, rather than mere computerization in local offices." Therefore, the mobile government is one of the most important E-Government developments (Kesavarapu and Choi, 2012) and the development of government mobile applications will be booming in the coming years, helped by artificial intelligence techniques. Nowadays governments around the world are interested in the implementation of artificial intelligence, thinking about the strategies, opportunities, and risk related to the implementation of these emerging technologies (Valle-Cruz et al., 2019). This whole range of technologies has the potential to benefit society, yet it is impossible to implement them in some areas with extreme poverty.

DIGITAL DIVIDE

The aim of this section is to explain what the digital divide is towards the connection between technology in government and the diffusion of innovations. The digital divide refers to the disparity between individuals, households, enterprises and geographical areas at different socio-economic levels concerning their opportunities to access information and communication technologies and use the Internet (OECD, 2001). It reflects differences between and within countries and raises several issues that can be explained from different points of view, taking into account the adoption of technology.

Castells (1998) argues that the wealth generation, the exercise of power and the creation of cultural codes became dependent on the technological capacity of societies and individuals, with information technologies at the core of this capacity.

The process of diffusion of related innovations exhibits behavior similar to an S-curve, with the center representing development and the periphery representing underdevelopment (Mahajan and Peterson, 1985; Rogers, 2003). Such behavior creates a divide between those who can benefit first from innovation and those who are excluded, for the dissemination of ICTs it is called the digital divide. Conceptualizing digital divide Hilber (2010) argues that:

[…] distinctions can be made in terms of the user group; the type of technology being considered and the stage of adoption. Some notions of digital divide select a specific technological solution as a representation of the bulk of digital technologies (such as telephone or Internet subscription) and compare the amount of equipment or services between societies (international digital divide) or within different social segments of society (national digital divide). In addition, different stages in the process of technology adaptation can be distinguished. Rogers (2003) has distinguished five different stages of adoption. Statisticians interested in measuring the nature of the digital divide have merged these five stages into three consecutive steps: Access, use, and impact of ICTs […] (p. 4).

In general terms, digital divide is a challenge of developing countries due to the deployment of technology has been very unequally distributed (Mariscal, 2005). In this paper we analyze how digital divide in Mexico is evolving and changing in different regions of the country.

TECHNOLOGY AND THE DIFFUSION OF THE INNOVATION THEORY

In this section, we describe, in detail, the diffusion of the innovation theory. Diffusion is the process by which an innovation communicates through certain channels over time among members of the social system (Koçak, Kaya, & Erol, 2013: p. 23). Diffusion is a special type of communication, where messages are related to a new idea. The diffusion of the innovation theory seeks to explain the adoption and spread of new ideas, products or services across different communication channels over time and in a particular social system (Bakshy, Karrer, & Adamic, 2009; Dodds & Watts, 2004; Rogers, 2003; Toole, Cha, & González, 2012; Valente, 1996). Adopters (e.g., individuals, organizations, states) are classified into different levels of disposition to accept or reject innovations, according to Koçak, Kaya, & Erol (2013):

[…] diffusion is also defined as the process by which an innovation is adopted and gained acceptance by members of a certain community […] (p. 23).

Members of each category have distinctive features based on the relative timing of innovation acceptance: innovators, early adopters, early majority, late majority, and laggards; diffusion is the process by which an innovation is communicated through certain channels over time among the members of a social system (Rogers 2003).

Although there is a gap related to the study of diffusion of innovations and government, a seminal work by Walker (1969), designed a score to classify the diffusion of innovations among the American states. Young (2006) studied the diffusion of innovation in social networks, considering processes in which new technologies and forms of behavior are transmitted through social or geographic networks. Some other research is related to the design of models that integrates the technology-organization-environment framework and four factors that are central to adoption decisions: perceived benefits, perceived barriers, organizational readiness, and external pressures, to examine factors influencing the adoption of open government data among government agencies in Taiwan. The results show a significant positive relationship among perceived benefits, organizational readiness, and external pressures and the adoption of open government data by government agencies (Wang and Lo, 2016).

Caiazza (2016) analyzed the role of policymakers in promoting new technology diffusion, to identify the barriers that affect the process of innovation diffusion and that are relevant for public policy-makers and to analyze potential policies to overcome the main barriers to the diffusion of new relevant technologies. Another study is related to the innovation in the public sector to design a future research agenda (Vries, Bekkers, and Tummers, 2016), and Aizstrauta and colleagues (2015) evaluated technologies that combine socio-economic aspects and socio-technical characteristics of technology development and exploitation using diffusion of innovations for the evaluation of the integrated acceptance and sustainability assessment model.

The diffusion of social media among public administrations has significantly grown. This phenomenon has created a field of research that seeks to understand adoption and impact of social media in the public sector (Criado, Sandoval-Almazan, & Gil-Garcia, 2013; Criado, Rojas-Martín, & Gil-García, 2017). People are so familiar with the use of Internet, that social media is the final step in the evaluation process of the Internet and can be considered as a great innovation (Koçak, Kaya, & Erol, 2013).

Internet-based applications have gained great popularity in the last ten years with millions of users. Because of the expansion and diversification of Internet applications, it becomes a part of individuals' daily lives (p. 25).

Factors that interact to influence the diffusion process are the innovation itself, how information about the innovation is communicated, time, and the nature of the social system to which the innovation is being introduced (Folorunso, Vincent, Adekoya, & Ogunde, 2010: p 362). According to Reich (2016):

[...] Agents, who are all using some old technology, choose whether or not to adopt the new one. We consider technologies with complementarities, so the usefulness of the new technology depends on who else is using it [...] the usefulness of a

communication technology is dependent on an individual's friends, family and other contacts using it. Agents choose to adopt the new technology only when a high enough proportion of their social contacts also adopt it. Agents can choose to adopt or not independently. They may also take joint decisions with others to adopt the new technology together. Once some agents adopt the technology, their contacts may adopt it; then their contacts may adopt it too, and so on [...] (p. 8).

The diffusion of innovations theory has guided multiple studies of the adoption of social media by state governments. In a study of the use of social networks sites (SNS) in 75 of the largest US cities between 2009 and 2011, Mossberger, Wu, and Crawford (2013) found that the adoption of Facebook skyrocketed from just 13% of the cities in 2009 to nearly 87% in 2011. Similarly, the use of Twitter by these cities increased from 25% to 87%. The authors did not study the causes of these increases in SNS and Twitter adoption. In a study of US federal agencies using interviews with social media directors, Mergel (2013) identified three key social media adoption tactics for federal government agencies: (1) representation, (2) engagement and (3) networking.

Montanari & Saberi (2010) found that innovation in social networks spreads much slower on well-connected network structures dominated by long-range links than in low-dimensional ones dominated, for example, by geographic proximity. Anderson, Lewis, & Dedehayir (2015) argue that there are few empirical studies on the diffusion of technological innovations across the public sector.

In the diffusion of innovations theory, members of each category have distinctive features based on the relative timing of innovation acceptance: innovators, early adopters, early majority, late majority, and laggards, which can be summarized as follows (Rogers 2003).

Innovators

According to Rogers (2003), innovators represent about 2.5% of the total population; they assume risks introducing and spreading innovations. They are mainly the producers of innovations, being helped by enthusiastic people that can influence others.

Early Adopters

Represent about 13.5% of the total population; they adopt innovation for the first time without much discussion or analysis. They can be recognized and respected as leaders, and they can persuade others to adopt innovation.

Early Majority

Represents about 34% of the total population; they are unwilling to take risks and are resistant to changes. They analyze their decisions. However, they are likely to accept innovation.

Late Majority

Represents about 34% of the total population; they characterize because of the resistance to changes and are difficult to persuade in adopting innovations without an intensive activity and significant influence.

Laggards

Represents about 16% of the total population; this is the category where people are more reluctant to changes; they are indifferent to any innovation and can even oppose it and fight it actively. They are jealous guardians of the status quo and frequently never adopt innovations.

Rogers (2003) graphed his findings using the normal curve. When the adoption curve is viewed as a curve of percentage, it takes the shape of an "S" curve, representing the adoption rate of innovation in a population. In the next stage, we present different studies related to social media and government.

This paper, specifically, tries to contribute in this area.

METHODS

This research is part of a longitudinal study of technology in Mexico for which we collected data of all the Mexican state governments. For the study reported here, we collected data from the National Ministry of Statistics and Data (INEGI) from all 32 Mexican state governments, from 2010 through 2018 and conducted state analyses to determine their technology adoption based on the Rogers' theory.

Our research methods have four main stages. For the first stage, we collected secondary data of all 32 Mexican state governments. Second, we compared annual datasets to find the common variables related to information technology between 2010 and 2018. The variables we analyzed were related to the Module on Availability and Use of Information Technologies in Households 2010 – 2014 (MODUTIH) and the National Survey on Availability and Use of Information Technologies in Households 2105 – 2018 (ENDUTIH), see Table 1.

Table 1. Percentage of households with information and communications technology equipment 2010 - 2018

Year	Computer	Internet	Television	Telephone (land line)	Radio	Electricity
2010	29.8	22.2	94.7	80.6	82.5	99.3
2011	30.0	23.3	94.7	82.2	81.0	99.2
2012	32.2	26.0	94.9	83.6	79.3	99.2
2013	35.8	30.7	94.9	85.5	76.9	99.3
2014	38.3	34.4	94.9	87.4	73.3	99.5
2015	44.9	39.2	93.5	89.3	65.8	99.2
2016	45.6	47.0	93.1	90.1	61.5	99.3
2017	45.4	50.9	93.2	91.9	58.6	99.4
2018	44.9	52.9	92.9	92.2	56.2	99.3

Although the methodology changes between each type of secondary source, the methodology we provide in this document, based on data standardization, allowed us to generate a final measure of innovation in the counties of Mexico. However, the disaggregated data by state and type of technology for 2012 were not found in the INEGI system. This situation did not prevent us from carrying out the analysis we had planned. We analyze the behavior of variables trying to identify the increasing adoption of technology and the "S" curve proposed by Rogers (Figure 1).

The variables that do not show growth in the percentage of adoption are television, radio, and electricity. These are traditional technologies and do not represent innovation for the Mexican context. However, the computer, Internet and even the telephone show growth in their adoption in the percentage of the population of each state government. For this reason, only the variables percentage of households with a computer, telephone, and Internet by state were used to carry out the analysis, these variables allowed us to identify the behavior of increasing use of technology in each state government.

Mass media such as television and radio have declined in adoption because information technologies such as the Internet and computers have replaced or absorbed them, nowadays there are online streaming plans services like Netflix and Amazon, or for playing music such as Spotify and Google Play Music, even radio stations are streaming live over the Internet; The Internet is the basis for implementing social media based services and E-Government.

For the third stage, we calculate normalized values of each state and each year, in order to determine how long each state government has used technology, using

Figure 1. Polynomial behavior of technology variables

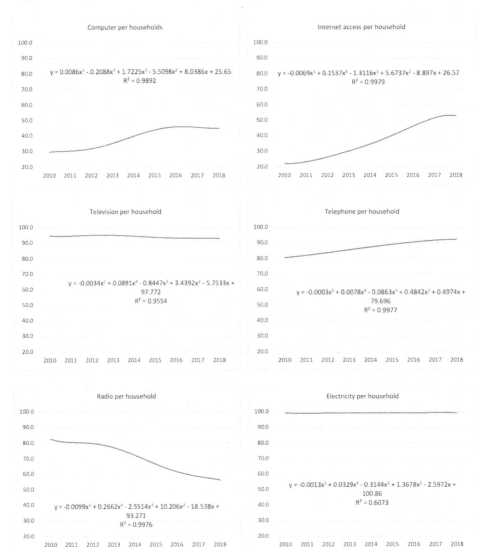

the normal form $\left(z = \dfrac{x - \mu}{\sigma} \right)$ $z = \dfrac{x - \frac{1}{4}}{\tilde{A}}$ for for each value to homogenize units of each variable. This way we can calculate a total average, of each county, that we called *Rank*. Later we calculated an average value based on the normalized variables (Table 2); we calculated a simple average (*Rank*). The *"Rank"* value allowed us to classify the technology adoption of governments from 2010 to 2018, considering percentage of households with a computer, telephone, and Internet.

Table 2. Ranking of DIFUSSION of innovations for Mexican State governments

State governments (counties)	Rank (2010-2018)	Classification based on the Diffusion of Innovations Theory
Mexico City	1.578	Innovator
Baja California	1.402	Early adopter
Sonora	1.295	Early adopter
Baja California Sur	1.230	Early adopter
Nuevo León	1.201	Early adopter
Quintana Roo	0.829	Early majority
Colima	0.726	Early majority
Jalisco	0.713	Early majority
Aguascalientes	0.663	Early majority
Sinaloa	0.506	Early majority
Tamaulipas	0.440	Early majority
Chihuahua	0.437	Early majority
Coahuila de Zaragoza	0.329	Early majority
Morelos	0.293	Early majority
Estado de México	0.257	Early majority
Querétaro	0.144	Early majority
Yucatán	0.111	Late majority
Nayarit	0.082	Late majority
Durango	-0.074	Late majority
Campeche	-0.180	Late majority
Tabasco	-0.342	Late majority
Guanajuato	-0.406	Late majority
San Luis Potosí	-0.595	Late majority
Zacatecas	-0.634	Late majority
Michoacán de Ocampo	-0.683	Late majority
Hidalgo	-0.798	Late majority
Puebla	-0.854	Late majority
Tlaxcala	-0.995	Laggard
Veracruz	-1.049	Laggard
Guerrero	-1.565	Laggard
Oaxaca	-1.921	Laggard
Chiapas	-2.142	Laggard

Source: Own elaboration

FINDINGS

Based on the diffusion of innovations theory's adopters, we found the following results: 1 innovator, 4 early adopters, 11 early majority, 11 late majority, and 5 laggards. This distribution is very close to the predicted percentages. Our data show: 1 innovator out of 32 cases or 3.1% (2.5% predicted); 4 early adopters out of 32 cases or 12.5% (13.5% predicted); 11 early majority out of 32 or 34.4% (34% predicted); 11 late majority out of 32 cases or 34.4% (34% predicted); and 5 laggards out of 32 or 15.6% (16% predicted). We classified state governments depending on their final rank and classification, representing the technology adoption.

Our state government data analysis shows the trends of technology adoption in Mexico. By this measure, and according to the diffusion of innovations theory, Mexico City is the innovator. Baja California, Sonora, Baja California Sur, and Nuevo León are early adopters, and Tlaxcala, Veracruz, Guerrero, Oaxaca, and Chiapas are laggards; The northern states are early adopters and early majority; most of the states in the north central part are late majority, and the states of the southern part are laggards (Figure 2).

Figure 2.
Source: Own elaboration

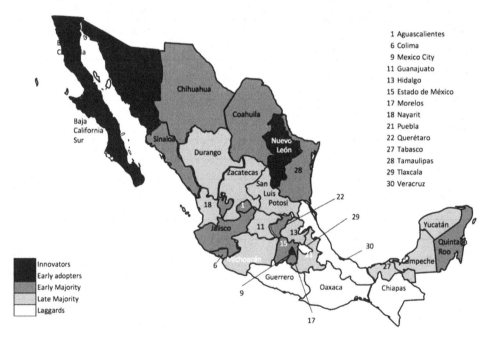

Socio-economic data of the states and their populations are important in this study, because it may help to explain technology adoption in the country. Early adopters of innovations tend to be better educated and have a higher income, even within different social strata. Data shown in Table 3 were obtained from the National Ministry of Statistics and Data (INEGI) and show the population size, illiteracy; the contribution percentage to the National Gross Domestic Product (GDP), and the access to social network sites are shown for 2014, since this is the most recent year that INEGI presented these data.

Mexico City, classified as an innovator, being the county with the highest percentage of contribution to the National Gross Domestic Product (GDP). Estado de México is the second highest contributor (classified as an early majority); Nuevo León (an early adopter) is the third highest contributor to GDP, Colima (early majority), and Tlaxcala (Laggard) have the lowest contribution percentage to the National GDP.

Nuevo León, an early majority in our classification, has a high percentage level of contribution to the National GDP and households with a computer, with the second lowest level of illiteracy. However, population size does not seem to be an important factor in this case. For this data, we could not find a relationship between innovation adoption and socio-economic level.

Estado de México and Mexico City are the most populated states; Baja California Sur (early adopter) and Colima (early majority) have the smallest population. Mexico City (the innovator), Nuevo Leon and Baja California (early adopters) have the lowest levels of illiteracy, and Chiapas, Guerrero, Oaxaca, and Veracruz have the highest levels of illiteracy.

The two most populated states are Estado de México and Mexico City; these states are in the early majority and the innovator. However, population size is not a strong factor to define innovation's adoption, because Baja California Sur is an early adopter with a relatively small population as well as a reduced GDP percentage contribution.

The adoption of technology by state governments needs more research on different fronts. Even the adoption of disruptive technologies such as artificial intelligence will need to be analyzed. Many governments – state or federal – use these technologies without many strategies about how to increase their value. Only a few of them know about the effects of this interaction with citizens and their use because they lack methods and metrics to understand it. The purpose of this research is to provide some direction about this problem using the diffusion of the innovation theory.

The most influential variable in the adoption of information technologies is illiteracy, and there is an inverse relationship between technology and illiteracy, the second most important, but with a low correlation is GDP (Table 4).

Table 3. Socio-economic data of Mexican state governments

State government	Rank	Classification	% GDP (2014)	Population (2015)	Illiteracy (2015)	Access to SNS (2014)
Mexico City	1.578	Innovator	16.5	8,918,653.00	1.5	33%
Baja California	1.402	Early adopter	2.8	3,315,766.00	2	54%
Sonora	1.295	Early adopter	2.9	2,850,330.00	2.2	47%
Baja California Sur	1.230	Early adopter	0.7	712,029.00	2.5	33%
Nuevo León	1.201	Early adopter	7.3	5,119,504.00	1.6	42%
Quintana Roo	0.829	Early majority	1.6	1,501,562.00	3.9	42%
Colima	0.726	Early majority	0.6	711,235.00	3.9	50%
Jalisco	0.713	Early majority	6.5	7,844,830.00	3.5	40%
Aguascalientes	0.663	Early majority	1.2	1,312,544.00	2.6	38%
Sinaloa	0.506	Early majority	2.1	2,966,321.00	4.2	46%
Tamaulipas	0.440	Early majority	3	3,441,698.00	3	52%
Chihuahua	0.437	Early majority	2.8	3,556,574.00	2.6	55%
Coahuila de Zaragoza	0.329	Early majority	3.4	2,954,915.00	2	27%
Morelos	0.293	Early majority	1.2	1,903,811.00	5	38%
Estado de México	0.257	Early majority	9.3	16,187,608.00	3.3	47%
Querétaro	0.144	Early majority	2.2	2,038,372.00	4.5	43%
Yucatán	0.111	Late majority	1.5	2,097,175.00	7.4	39%
Nayarit	0.082	Late majority	0.7	1,181,050.00	5	48%
Durango	-0.074	Late majority	1.2	1,754,754.00	3.2	35%
Campeche	-0.180	Late majority	4.2	899,931.00	6.6	26%
Tabasco	-0.342	Late majority	3.1	2,395,272.00	5.4	38%
Guanajuato	-0.406	Late majority	4.2	5,853,677.00	6.3	42%
San Luis Potosí	-0.595	Late majority	1.9	2,717,820.00	6.3	34%
Zacatecas	-0.634	Late majority	1	1,579,209.00	4.4	50%
Michoacán	-0.683	Late majority	2.4	4,584,471.00	8.3	47%
Hidalgo	-0.798	Late majority	1.7	2,858,359.00	8.3	38%
Puebla	-0.854	Late majority	3.2	6,168,883.00	8.3	36%
Tlaxcala	-0.995	Laggard	0.6	1,272,847.00	3.9	28%
Veracruz	-1.049	Laggard	5.1	8,112,505.00	9.4	32%
Guerrero	-1.565	Laggard	1.5	3,533,251.00	13.6	54%
Oaxaca	-1.921	Laggard	1.6	3,967,889.00	13.3	34%
Chiapas	-2.142	Laggard	1.8	5,217,908.00	14.8	38%

Source: Self-elaboration with references in INEGI (2014a, 2014b, 2015a, 2015b, 2015c)

Table 4. Pearson's correlation coefficient

Person correlation	% GDP (2014)	Population (2015)	Illiteracy (2015)	Access to SNS (2014)
Rank	0.35	-0.03	-0.88	0.20

CONCLUSION

In this research, we suggested a behavior based on the diffusion of the innovation theory (Rogers, 2003), explaining the way state governments adopt technology, classifying it into five categories: innovators, early adopters, early majority, late majority, and laggards. In a threshold: from the risk takers to those who are actively against change, the average value (rank) we calculated represents this scale, from the innovator with the biggest average value to laggards with the most negative value.

The classification of innovation adopters suggests some general observations. The first one is that adoption of technologies by Mexican states, except for Mexico City and Quintana Roo, technology adoption is concentrated in the north of the country, perhaps across the U.S. border. The counties to the south of the country are the laggards, and their geographical location may make it difficult to implement technologies, as well as culture and illiteracy levels. We found that more illiteracy, less adoption of technologies, which means that people with less education level have less access to technology (Mahajan and Peterson, 1985; Rogers, 2003); but we also infer an inverse effect, that due to the lack of services and technologies illiteracy is not eradicated. This is manifested in the areas with the greatest poverty in the country and the biggest digital divide (Hilber, 2010).

For the early majority component, we have a combination of different state governments related to technology adoption. For example, northern border states like Chihuahua, Coahuila, and Tamaulipas; Quintana Roo in southeastern Mexico; and Aguascalientes, Colima, and Jalisco in the center are included in this category. However, Baja California, Sonora, and Nuevo León are located on the border with the United States and are early adopters, perhaps because they are heavily involved in communication between migrants and customs activities with the US border. Furthermore, the capital cities of Baja California (Tijuana) and Nuevo Leon (Monterrey) are two of the most important cities in Mexico, based on diverse industrial activities. Quintana Roo is in the southeast is a place that has many tourism activities; for example, Cancun is one of the most visited cities in Mexico, by Mexicans and foreigners. On the contrary, although Veracruz has the largest international seaport in the country, and Guerrero is a popular place for tourism activities, they are among the laggards regarding technology adoption.

Some of the state governments in the late majority and laggard categories have less economic development and populations with less education and income. For example, states such as Oaxaca, Chiapas, and Hidalgo are less developed and have a lower overall technology adoption.

The population of the state of Colima, an early adopter, has high illiteracy and more labor-intensive industries, such as agriculture, rather than more information-based economic activities. However, Baja California Sur with a similar population and GDP has less illiteracy and is an early majority.

The differences between innovators 1.578 and the most laggards 2.142 is also evident in Mexico deserving further study in order to understand the limitations and challenges that these state governments face in the adoption of technology. The earliest adopters of technology in Mexico begins in the center of the country and ends in the north.

We expect that data from ten years of quantitative data of technology – households with a computer, telephone, and Internet by state – provide important insights into the adoption of technology by Mexican state governments.

In this paper, we propose a method to classify data related to technology adoption based on diffusion of innovations theory and provide a statistical analysis that allows us to aggregate and normalize these data.

The diffusion of the innovation theory provides a systematic approach to understand where the efforts of governments are in this adoption process. We have shown that Mexican state governments' adoption of technology is increasing and tends to increase coverage in the counties. Furthermore, we provide empirical evidence that mass media as television and radio is decreasing in use, being replaced by the Internet and telephony. We have proposed a categorization method using a diffusion of innovation concepts that should be helpful with future research on emerging technologies adoption by governments. Although we did not study the strategies, political and institutional variables, the paper provides empirical evidence for the governments of lower-ranking, useful to replicate the technology strategies of the innovator and early adopters.

There are important limitations to the research presented here. As mentioned earlier, the speed of change of these technologies along with the increase in the number of new users will bias this research and provide other results. Also, the state governments strategies that promote the use of technologies can alter these results in the short term, and we suggest that a combination of quantitative and qualitative research must be done in order to have an integrated way to see the problem holistically.

Future research will open several paths for understanding different adoption behaviors among specific technologies in each state, such as big data, artificial intelligence, Internet of things, and smart cities. A more contextualized approach could lead to new explanations of the phenomena more closely related to different technical issues.

REFERENCES

Aizstrauta, D., Ginters, E., & Eroles, M. A. P. (2015). Applying theory of diffusion of innovations to evaluate technology acceptance and sustainability. *Procedia Computer Science*, *43*, 69–77. doi:10.1016/j.procs.2014.12.010

Anderson, M., Lewis, K., & Dedehayir, O. (2015). Diffusion of innovation in the public sector: Twitter adoption by municipal police departments in the U.S. In *International Conference on Management of Engineering and Technology (PICMET)*. Portland, OR: IEEE. 10.1109/PICMET.2015.7273207

Arias Torres, D., & Manriquez, J. C. (2017). Evolución del E-Gobierno 1.0 al 4.0. *U-GOB*. Available at: https://www.u-gob.com/evolucion-del-e-gobierno-1-0-al-4-0/

Au, M., Lam, J., & Chan, R. (2015). Social Media Education: Barriers and Critical Issues. *Communications in Computer and Information Science*. doi:10.1007/978-3-662-46158-7_20

Bakshy, E., Karrer, B., & Adamic, L. A. (2009). Social Influence and the Diffusion of User-created Content. In *Proceedings of the 10th ACM Conference on Electronic Commerce* (pp. 325–334). New York, NY: ACM. 10.1145/1566374.1566421

Banerjee, A., & Jain, S. (2003). *e-Governance in India: Models That Can Be Applied in Other Developing Countries*. Available at: http://www.springerlink.com/openurl.asp?genre=article&id=KHV53J4VQHHLNC4X

Brown, M. M., & Brudney, J. L. (2004). Achieving Advanced Electronic Government Services: Opposing Environmental Constraints. *Public Performance & Management Review*, *28*(1), 96–113.

Carroll, J. (2005). *Risky Business: Will Citizens Accept M-government in the Long Term?* Mobile Government Consortium International LLC. Available at https://www.researchgate.net/publication/228847573_Risky_Business_Will_Citizens_Accept_M-government_in_the_Long_Term%27

Chatfield, A. T., & Reddick, C. G. (2018). The role of policy entrepreneurs in open government data policy innovation diffusion: An analysis of Australian Federal and State Governments. *Government Information Quarterly, 35*(1), 123–134. doi:10.1016/j.giq.2017.10.004

Criado, J. I., Rojas-Martín, F., & Gil-García, J. R. (2017). Enacting social media success in local public administrations: An empirical analysis of organizational, institutional, and contextual factors. *International Journal of Public Sector Management, 30*(1), 31–47. doi:10.1108/IJPSM-03-2016-0053

Criado, J. I., Sandoval-Almazán, R., & Gil-García, J. R. (2013). Government innovation through social media. *Government Information Quarterly, 30*(4), 319–326. doi:10.1016/j.giq.2013.10.003

De Vries, H., Bekkers, V., & Tummers, L. (2016). Innovation in the public sector: A systematic review and future research agenda. *Public Administration, 94*(1), 146–166. doi:10.1111/padm.12209

Dodds, P. S., & Watts, D. J. (2004). Universal Behavior in a Generalized Model of Contagion. *Physical Review Letters, 92*(21), 218701. doi:10.1103/PhysRevLett.92.218701 PMID:15245323

Emmanouilidou, M., & Kreps, D. (2010). A framework for accessible m-government implementation. *Electronic Government, an International Journal, 7*(3), 252–269.

Folorunso, O., Vincent, R. O., Adekoya, A. F., & Ogunde, A. O. (2010). Diffusion of innovation in social networking sites among university students. *International Journal of Computer Science and Security, 4*(3), 361–372.

Gil-García, J. R. (2012). Towards a Smart State? Inter-agency collaboration, information integration, and beyond. *Information Polity, 17*(3), 269–280. doi:10.3233/IP-2012-000287

Gil-García, J. R., Dawes, S. S., & Pardo, T. A. (2018). *Digital government and public management research: finding the crossroads.* Academic Press.

Hilbert, M. (2010). When is cheap, cheap enough to bridge the digital divide? Modeling income related structural challenges of technology diffusion in Latin America. *World Development, 38*(5), 756–770. doi:10.1016/j.worlddev.2009.11.019

INEGI. (2014a). *Aportación al Producto Interno Bruto (PIB) nacional.* Instituto Nacional de Estadística y Geografía. Online: http://www.cuentame.inegi.org.mx/monografias/informacion/df/economia/pib.aspx?tema=me&e=09

INEGI. (2014b). *Usuarios de Internet por tipo de uso según entidad federativa, 2014*. Institito Nacional de Estadística y Geografía. Online: http://www3.inegi.org.mx/sistemas/sisept/default.aspx?t=tinf255&s=est&c=28978

INEGI. (2015a). *Analfabetismo*. Instituto Nacional de Estadística y Geografía. Online: http://cuentame.inegi.org.mx/poblacion/analfabeta.aspx?tema=P

INEGI. (2015b). *Encuesta Nacional sobre la Disponibilidad y Uso de Tecnologías de la Información en los Hogares (ENDUTIH) 2015*. Instituto Nacional de Estadística y Geografía. Online: https://www.inegi.org.mx/programas/modutih/2014/

INEGI. (2015c). *Número de habitantes*. Instituto Nacional de Estadística y Geografía. Online: http://cuentame.inegi.org.mx/poblacion/habitantes.aspx?tema=P

Karo, E., & Kattel, R. (2019). Public Administration, Technology and Innovation: Government as Technology Maker? In Public Administration in Europe (pp. 267-279). Palgrave Macmillan.

Kesavarapu, S., & Choi, M. (2012). M-government - a framework to investigate killer applications for developing countries: An Indian case study. *Electronic Government. International Journal (Toronto, Ont.)*, *9*(2), 200–219.

Koçak, N. G., Kaya, S., & Erol, E. (2013). Social Media from the Perspective of Diffusion of Innovation Approach. *The Macrotheme Review*, *2*(3), 22–29.

Kushchu, I. (2007). *Mobile Government: An Emerging Direction in E-government* (1st ed.). Hershey, PA: IGI Publishing. doi:10.4018/978-1-59140-884-0

Lu, M. (2001). Digital Divide in Developing Countries. *Journal of Global Information Technology Management*, *4*(3), 1–4. doi:10.1080/1097198X.2001.10856304

Luna-Reyes, L. F., García, H., Manuel, J., & Gil-García, J. R. (2009). Hacia un modelo de los determinantes de éxito de los portales de gobierno estatal en México. *Gestión y Política Pública*, *18*(2), 307–340.

Magnusson, M., Bellström, P., & Thoren, C. (2012). *Facebook usage in government – a case study of information content*. Available at: https://aisel.aisnet.org/amcis2012/proceedings/EGovernment/11

Mahajan, V., & Peterson, R. A. (1985). *Models for Innovation Diffusion*. Sage Publications. doi:10.4135/9781412985093

Mariscal, J. (2005). Digital divide in a developing country. *Telecommunications Policy*, *29*(5-6), 409–428. doi:10.1016/j.telpol.2005.03.004

Mergel, I. (2013). Social media adoption and resulting tactics in the U.S. federal government. *Government Information Quarterly*, *30*(2), 123–130. doi:10.1016/j.giq.2012.12.004

Montanari, A., & Saberi, A. (2010). The spread of innovations in social networks. *Proceedings of the National Academy of Sciences of the United States of America*, *107*(47), 20196–20201. doi:10.1073/pnas.1004098107 PMID:21076030

Montiel, M., & Manuel, J. (2016). The digital divide in Mexico: A mirror of poverty. *Mexican Law Review*, *9*(1), 93–102. doi:10.1016/j.mexlaw.2016.09.005

Mossberger, K., Wu, Y., & Crawford, J. (2013). Connecting citizens and state governments? Social media and interactivity in major U.S. cities. *Government Information Quarterly*, *30*(4), 351–358. doi:10.1016/j.giq.2013.05.016

OECD. (2001). *Understanding The Digital Divide*. Paris: OECD. Available at: http://www.oecd.org/dataoecd/38/57/1888451.pdf

Palkovits, S., Woitsch, R., & Karagiannis, D. (2003). Process-Based Knowledge Management and Modelling in E-government — An Inevitable Combination. In M. A. Wimmer (Ed.), *Springer Berlin Heidelberg* (pp. 213–218). doi:10.1007/3-540-44836-5_22

Reich, B. (2016). *The diffusion of innovations in social networks*. London: University College London.

Reinermann, H. L. J. von (2000). *Speyerer Definition von Electronic Government*. Available at: http://www.joernvonlucke.de/ruvii/Sp-EGov.pdf

Rogers, E. M. (2003). *Diffusion of Innovations* (5th ed.). New York, NY: Simon and Schuster.

Sandoval-Almazán, R., Cruz, D. V., & Armas, J. C. N. (2015, January). Social media in smart cities: An exploratory research in Mexican municipalities. In *2015 48th Hawaii International Conference on System Sciences* (pp. 2366-2374). IEEE.

Sandoval-Almazán, R., & Gil-García, J. R. (2011). The Role of the CIO in a Local Government IT Strategy: The case of Merida, Yucatán Mexico. *Electronic Journal of E-Government*, *9*(1), 1–14.

Sandoval Almazán, R., & Gil García, J. R. (2013). Social Media in State Governments: Preliminary Results About the Use of Facebook and Twitter in México. In Z. Mahmood (Ed.), *E-Government Implementation and Practice in Developing Countries* (Vol. 1, pp. 128–146). Hershey, PA: IGI Global. doi:10.4018/978-1-4666-4090-0.ch006

Sandoval-Almazán, R., Gil-García, J. R., Luna-Reyes, L. F., Luna, D. E., & Rojas-Romero, Y. (2012, October). Open government 2.0: citizen empowerment through open data, web and mobile apps. In *Proceedings of the 6th International Conference on Theory and Practice of Electronic Governance* (pp. 30-33). ACM. 10.1145/2463728.2463735

Sandoval-Almazán, R., & Valle-Cruz, D. (2016). Social Media in Local Governments in Mexico: A Diffusion Innovation Trend and Lessons. In *Social Media and Local Governments* (pp. 95–112). Cham: Springer. doi:10.1007/978-3-319-17722-9_6

Sandoval-Almazán, R., Valle-Cruz, D., & Kavanaugh, A. L. (2018). The diffusion of social media among state governments in Mexico. *International Journal of Public Administration in the Digital Age*, 5(1), 63–81. doi:10.4018/IJPADA.2018010104

Scheider, D. M. G. (2000). *Meta-capitalism: The e-business revolution and the design of 21st century companies and markets*. New York: John Wiley & Sons Inc.

Scholl, H. J. (2010). *Electronic Government: Information, Technology, and Transformation*. Armonk, NY: ME Sharpe.

SENER. (2017). *Regiones sin electricidad. Secretaría de Energía*. Online: https://datos.gob.mx/busca/dataset/regiones-sin-electricidad

Sundar, D. K., & Garg, S. (2005). *M-Governance: A Framework for Indian Urban Local Bodies*. Mobile Government Consortium International LLC.

Toole, J. L., Cha, M., & González, M. C. (2012). Modeling the Adoption of Innovations in the Presence of Geographic and Media Influences. *PLoS One*, 7(1), e29528. doi:10.1371/journal.pone.0029528 PMID:22276119

Tuten, T., Wetsch, L., & Munoz, C. (2015). Conversation Beyond the Classroom: Social Media and Marketing Education. *Developments in Marketing Science: Proceedings of the Academy of Marketing Science*, 317–317. 10.1007/978-3-319-11797-3_182

Valente, T. W. (1996). Social network thresholds in the diffusion of innovations. *Social Networks*, 18(1), 69–89. doi:10.1016/0378-8733(95)00256-1

Valle-Cruz, D. (2019). Public value of e-government services through emerging technologies. *International Journal of Public Sector Management*, 32(5), 530–545. doi:10.1108/IJPSM-03-2018-0072

Valle-Cruz, D., Ruvalcaba-Gomez, E. A., Sandoval-Almazán, R., & Criado, J. I. (2019). A Review of Artificial Intelligence in Government and its Potential from a Public Policy Perspective. In *20th Annual International Conference on Digital Government Research*. ACM. DOI: 10.1145/3325112.3325242

Valle-Cruz, D., Sandoval-Almazán, R., & Gil-García, J. R. (2016). Citizens' perceptions of the impact of information technology use on transparency, efficiency and corruption in local governments. *Information Polity, 21*(3), 1–14.

Walker, J. L. (1969). The diffusion of innovations among the American states. *The American Political Science Review, 63*(3), 880–899. doi:10.1017/S0003055400258644

Wang, H. J., & Lo, J. (2016). Adoption of open government data among government agencies. *Government Information Quarterly, 33*(1), 80–88. doi:10.1016/j.giq.2015.11.004

Wu, J., & Zhang, P. (2018). Local government innovation diffusion in China: An event history analysis of a performance-based reform program. *International Review of Administrative Sciences, 84*(1), 63–81. doi:10.1177/0020852315596211

Yildiz, M. (2007). E-government research: Reviewing the literature, limitations, and ways forward. *Government Information Quarterly, 24*(3), 646–665. doi:10.1016/j.giq.2007.01.002

Young, H. P. 2006). The diffusion of innovations in social networks. *The Economy as an Evolving Complex System III: Current Perspectives and Future Directions*, 267.

Chapter 4

Strategies for Managing E–Records for Good Governance:
Reflection on E–Government in the Kingdom of Eswatini

Vusi Tsabedze
https://orcid.org/0000-0001-9223-4266
University of South Africa, South Africa

ABSTRACT

Management of e-records has become an exponential factor that requires adequate consideration and planning in this era of digital technology. The use of e-records becomes significant such that e-government must implement its management for good governance in the public sector. As government of Eswatini is pursuing strategies to implement e-government, strategies to enhance the effectiveness of e-government programmes and operation becomes essential. This would help promote transparency, accountability, and good governance using information and communication technologies. The objective of this chapter is to determine infrastructure and strategies for managing e-records in an e-government context, to determine the risks of managing e-records as a strategic resource, and lastly, to look at prospects of e-records management in Eswatini. The chapter reviews the situation in Eswatini, drawing from other cases in the world.

DOI: 10.4018/978-1-7998-1526-6.ch004

INTRODUCTION

As governments embark on e-government, there is need to pay special attention to the management of electronic records (e-records). This is so because electronic transactions carried out through e-government applications produce e-records whose quality and integrity need to be upheld (IRMT, 2004; Mnjama &Wamukoya, 2004). The IRMT (2004:1) thus cautions that, "funds and effort will likely be wasted unless e-government initiatives are supported by a solid records and information management programme." Taking this notion into account, it can be said that e-government can be successful if it is driven by a robust e-records management system.

The Commonwealth Secretariat (2013) argues that the major challenges facing the implementation of e-government in Eswatini and other Sub-Saharan African countries is the lack of a proper ICT infrastructure that support e-records management. The Commonwealth Secretariat (2013) is of the view that among other salient factors e-government can only be implemented successfully if it is supported by functional and readily accessible e-records.

E-records are information that are created by use of electronic technologies (Ambira, 2016). They are stored on various magnetic and optical storage devices and are products of computer hardware and software. E-records readiness on the other hand can be defined as the depth and breadth or the capacity of organisations in having the required institutional, legal framework, ICT infrastructure anchored on a systematic records and information management programme (Kalusopa, 2011). It is the depth and breadth or the capacity of organisations in having the required institutional, legal framework, ICT infrastructure, and, records and information management programme based on the generic information and recordkeeping practices (IRMT, 2004).

In this chapter e-record readiness is viewed as having a proper e-records management system that can support e-government and improved service delivery by government ministries to the citizens.

PROBLEM STATEMENT

The overarching problem that instigated this study is that while there is abundant evidence of the Eswatini government's undoubted ICT platforms that provide accurate and faster communication through the use of e-applications to access government services, the status of e-record readiness of this implementation has not been fully ascertained. Several authorities on records management such as IRMT (2004; 2009) underscore the fact that though e-government services produce e-records that document government transactions and online activities, their extent of the application records management functionalities remain in contention.

The Eswatini ICT legislative and policy framework of 2007 allows for the establishment of the e-government portal that should provide ubiquitous access and sharing of information through internet among government departments; yet there have been several instances where records captured and stored in the e-records system have been lost or could not be accessed by the user community. This implies that the drive in the implementation of the national e-government strategy is fraught, among other issues, with e-records management challenges of admissibility, authenticity and reliability which are a cornerstone for evidence in the administration of the state and general governance of the country (Tsabedze, 2011).

Studies elsewhere in Africa such as the IRMT (2003), Wamukoya and Mutula (2005); Moloi (2006); Nengomasha (2009) and Kalusopa (2011) all contend and underscore the need for a thorough e-records readiness as key to the implementation of e-records management programmes and ultimately e-government in the public sector. However, past studies in Eswatini show no research evidence that ascertain the depth of e-records readiness in the context of the current e-government strategy. Studies that have been documented on records management systems in the country have largely focused on paper-based records management in government ministries, such as one conducted by Tsabedze (2011). Specifically on e-records, the study by Ginindza (2008) attempted to study the general state of e-government in Eswatini in government ministries and departments. Others such as Maseko (2010) examined the management of audio-visual records at the Eswatini Television Authority (STVA). The Eswatini National Archives Report (2015) also have noted that the lack of comprehensive studies in Eswatini on e-records management has prejudiced the department and its partners in the Ministry of Information Technology that would be a blueprint that can guide the implementation of the e-records project. This has resulted in government ministries adopting an uncoordinated approach in managing e-records owing to the fact that both the ICT Policy and e-government 2013-2017 strategy are silent on how e-records management is supposed to be implemented in the face of the e-government drive. In the same vein, the Records and Archives Act of 1971 is also obsolete and therefore inadequate to address the issues of e-records management in the current digital era. There is currently a paucity of empirical studies which address e-records readiness with respect to e-Government, in the context of a developing country such as the Kingdom of Eswatini.

AIM AND OBJECTIVES OF THE STUDY

This study assesses e-records readiness in the Eswatini government ministries with a view to conceptualizing framework for the effective management of e-records as a facilitating tool for e-government.

The specific objectives of the study were to:

- establish the national legal and policy framework governing management of electronic records in government ministries in Eswatini in the context of e-government.
- ascertain the level of compliance to policies, standards, tools, procedures and responsibilities for e-records management in the government ministries.
- establish the e-records management products and technologies existing in the government ministries.
- examine resource capacity and training for e-records management staff in the government ministries.
- establish the depth of government wide digital preservation strategy in the government ministries.
- conceptualise a framework that may inform the appropriate management of e-records in the context of Eswatini e-government.

LITERATURE REVIEW

Several scholars contend that it is now almost impossible to study the outcome of e-government processes without referring to e-records in the digital environment. For example, Nolan (2001:188) regards the use of ICT systems as a dominant reform model for the public service when linking the implementation of ICT to effective documenting of government services and knowledge sharing. Good record keeping is thus essential for governments and public institutions at all stages of development; particularly so for developing countries. Poor record-keeping systems are a major barrier to institutional, legal and regulatory reform; anti-corruption strategies; poverty reduction and economic development (Lipchak and McDonald, 2003).

E-records are the by-products of e-government functions in which the information is represented in digital form, whether it is text, graphics, data, audio or images. Electronically- generated information provides crucial improvements in the efficiency and effectiveness of service delivery as citizens can interact with government agencies online without having to physically visit the offices (Kamatula, 2010: 152).

According to the International Records Management Trust (IRMT) (2004), e-records management programmes in most governments around the world have been motivated in part by the ongoing public sector reforms. Governments are recognizing the need to facilitate access to public services through e-Government. IRMT (2004) further noted that e-government has led to generation of vast quantities of e-records. Research in the management of e- records has received a lot of attention in developed countries with investigations focusing on practical solutions to the management of records (Keakopa, 2009).

Mnjama and Wamukoya (2006: 277) also state that the emergence of e-government has led to the creation of information which is in fact a valuable asset that must be managed and protected. In addition to providing essential evidence of organizational activities, transactions and decisions, e-records also support business functions and are essential for assessing organizational performance. Without reliable e-records, governments cannot manage state resources, revenues or public administration. Governments cannot provide services such as education and medical care. On the other hand, without accurate and reliable e-records and an effective system to manage them, governments cannot be held accountable for their decisions and actions, and the rights and obligations of citizens and legal entities cannot be maintained.

Management of e-Records and e-Government in Developed Countries

Xiaomi (2009) conducted a study to investigate the status of e-records management in the context of e-government in USA, New Zealand and UK to help support the development of ERM in China. The main aim of this study was to investigate how the United States, United Kingdom and New Zeeland are managing e- Records, implemented e-records systems and to determine what the corresponding implications in e-government were. The findings of the study were to assist the Chinese government in improving management of e-records in the context of e-Government. The study adopted documentary review as a data collection instrument where data was collected through documentary review of laws, policies and procedures from the three countries, for the management of e-records together with the e-Government. The study revealed that in all the three countries, management of e-records was successfully embedded in e-government strategies.

In the United States, e- records management was considered an important infrastructure for e-government and was part of 24 e-government initiatives (White House 2012, NARA 2006). In the United Kingdom, the management of e-records is incorporated into departmental e-commerce strategies as part of the business continuity plan, information risk management solutions and knowledge management initiatives in the e-government strategy. In New Zealand, the management of e-records was seen as part of the strategy for information and digital services for the public in the e-government strategy. The study found that in all three countries, e-records were managed as core national assets and resources. The study recommended that management of e-records is vital for effective e-government, so it is necessary to integrate e-records into the e-government strategy and to strengthen collaboration between e-government authorities and records management to get the benefits of e-government.

Kalcu (2009) investigated how Turkey had adopted record management practices with the increase of e-government services. The study sought to assess if there had been new approaches in management of e-records as the Turkey government moved significantly to e-Government. The government agencies were selected as a main sample of the study. The study adopted a qualitative approach where data was collected through interviews and literature review. The study revealed that in Turkey, the e-records management applications are developed within the framework of e-government and e-records management is considered to be significant in terms of overcoming the handicaps between the government and the citizen as well as cutting the red tape. These applications are thought to be contributory to the development of records management applications. However, it is a fact that the conditions of reliability and durability of the printed environment have not yet been achieved in the electronic environment (Kulcu, 2009). Within these conditions, carelessly taken steps would surely lead to a disaster. In a developing country like Turkey in particular, the relative cost reduction introduced by the virtual environment, the speed and efficiency may lead every administrator to prefer e-government and e-records management applications as an easy solution at the beginning. However, meeting the required legal and administrative criteria related to e-records management applications, filling the gaps, taking actions on the examples of the developed countries, and following the outputs of projects like InterPARES are considered to be quite significant (Kulcu, 2009).

Management of e-Records and e-Government in Africa

In a recent study Ambira (2016) investigated how management of e-records supported e-government in Kenya with a view to develop a best-practice framework for management of e-records in support of e-Government. The study investigated how management of e-records supported e-government in Kenya with a view to develop a best-practice framework for management of e-records in support of e-government. The study was underpinned by the European Commission's (2001) Model Requirements for e-records management (MoReq) and the United Nation's (2001) five-stage e-government maturity model as theoretical frameworks. The study adopted the interpretive research paradigm and qualitative approach. The study used face-face interviews as a data collection instrument. The study revealed that the general status of management of e-records in government ministries is inadequate to support e-government. The utilization of e-government in Kenya had grown significantly and more ministries were adopting e-government services. Although some initiatives have been undertaken to enhance management of e-records, the existing practices for management of e-records require development to ensure that they sufficiently support e-government. However, there are several challenges in the management of e-records that impact on implementation of e-government. The study concluded that the current practices for managing e-records in support of e-government implementation were not adequate. Ambira (2016) recommended a best-practise framework for managing e-records in support of e-government in Kenya.

An earlier empirical study by Nengomasha (2009) provides essential insights into the current study. Nengomasha (2009) aimed to answer the research question: "How can the e-records environment be strengthened to support e-government in Namibia?" The study used a qualitative approach through observations and interviews as data collection techniques. The study revealed that e-government in the Public Service of Namibia, is in the initial phase of implementation and has led to an increase in the creation of e-records. However, the status of records management in the Public Service of Namibia, which has a hybrid records system, that is, a paper and e-records environment, is very poor. This is evident in the officers' lack of understanding of what records are and the importance of records management; inadequate legal and regulatory environment; failure to follow laid down procedures and standards; absence of a records management disaster plan including digital preservation strategy; and inadequate resources, which includes lack of staff and skills to manage records in general and in particular, e-records. The Public Service of Namibia's score of 55 out of 120 in an e-records readiness assessment carried out as part of the study, signifies high risk, which means that government's e-records are at risk of misuse and loss. The study came to the conclusion that Namibia's e-government initiatives are not supported by a strong records management programme. This missing link needs to be rectified to ensure that Namibia benefits fully from its investment in e-Government.

The International Records Management Trust (IRMT) conducted a study in 2010-2011 to assess the status of records management in East Africa for support of e-government and freedom of information. The study focused on the five African countries Rwanda Kenya, Uganda, Burundi and Ethiopia. The aim of the study was to investigate the relationship "between records management and the current and planned directions for ICT/ e-government and FOI" and the extent to which records are capable of providing reliable evidence for governance.

The study found that across the region, governments are aggressively pursuing ICT and e-government initiatives are, to a greater or lesser extent, the same general path towards building FOI regimes. It also revealed the importance of records management, including the management of e-records in policy, capacity and position and the strength of records and archival authorities.

Key recommendations include the need for developing a digital preservation plan across the five countries. These will ensure that the preservation of e-records that is required is retained over a long-term. The study also provided reports on the five countries with different recommendations affecting each of the countries.

Nengomasha (2009) recommends an integrated records management programme for the public service of Namibia to improve the e- records environments. Such a programme would promote records management awareness; determine resource requirements; review the legal and regulatory framework; review records management standards and procedures; develop and maintain records centres; manage archives; implement an e-records management system; and ensure the sustainability of the programme through staff training and regular monitoring and evaluation. In view of the fact that the Public Service of Namibia might take the route of enterprise content management (ECM), and in recognition of the importance of inter-operability of information systems for information sharing, further investigation is required into the electronic information systems running in the public service and possibilities for their integration with an e-records management system, which the Office of the Prime Minister plans to roll out to the entire public service. Therefore, this study is useful to the current study, in that it assessed e- readiness based on selected existing IRMT indicators as this study does.

Mnjama and Wamukoya (2007) conducted a study on e-government and records management: an assessment tool for e-records readiness in government. In their study they indicated that with the proliferation of ICT, e-records are being created in the public sector, which caused challenges to management of such records by registry staff. The study adopted a qualitative approach where data was collected through literature review on records management, ICT and e-governance. In addition, the study also looks at challenges faced by registry staff in developing countries. The study revealed that developing countries have paper-based records management systems, while e-records management is scarce.

The key recommends establishing an e-record-setting tool to help countries, particularly in Africa assess their readiness for the adoption of e-records in an e-government environment. The tool provided twenty-one questions, which would be the criteria for assessing the government's readiness to manage e-records in support of e-government.

Moloi (2006) investigates the management of e-records in a government setting in Botswana. A two-stage research design strategy involving a case study of government ministries and a survey of the respondents within government ministries was used. The population of the study consisted of: Director, Botswana National Archives and Records Services (BNARS), a representative of the Director, the Department of Information Technology (DIT), records staff, IT specialists, and action officers. The findings showed that whereas e-records management in developed countries is receiving great attention, the same cannot be said of Botswana. E-records management within the government of Botswana is at infancy and is fairly new. Botswana lacks an e-records management policy, which makes it difficult to identify, maintain and preserve e-records. Key recommendations include the need for the Botswana government to consider among other things, benchmarking against best practices of developed countries with regard to the systematic management of e-records.

The above studies in developed countries and in Africa all contend and underscore the need for a thorough e-records readiness as key to the implementation of e-records management programmes and ultimately e-government in the public sector. The studies also show that e-records management strengthens e-government services by supporting business continuity, accountability and transparency, good governance, and evidence-based decision-making.

The Need for e-Records Readiness Assessments in the Context of e-Government in Eswatini

The implementation of e-government and the management of the proliferation of e-records in recent years have made it imperative that systems are put in place to capture records as evidence of business activities (Kennedy & Schauder, 1998). Since governments are now providing their services online, it is also essential that established e-records are reliable, authentic, usable, and integrated (ISO 15489-1: 2016).

The literature shows that, governments have acknowledged the close relationship between good governance, records management and expansion of the electronic world. As such, governments have intensified a framework of values, policies, standards, systems and individuals that allow the readiness of e-records. It has also emerged from the literature that proper records management helps to furnish accurate, timely, and complete information for efficient decision making in the management and operation of the organisation. Moreover, effective and efficient e-records are essential when it comes to evidence in cases of litigation. It has been noted, however, that, though in developed countries such as Britain, Canada, Australia and America e-records are given priority and standards and proper systems are in place; the same is not happening in developing countries.

The literature reviewed also revealed that most African countries continue to face several challenges in managing records, particularly e- records. Although most countries in ESARBICA have attempted to put in place some programmes to manage records in general, there are no known clear strategies initiated either to manage e-records or have e-records readiness assessments rigorously carried out (Kalusopa, 2011).

Keakopa (2010:67) in a recent critical appraisal of the management of e-records in the ESARBICA highlights the persistent "limitations from research conducted in the region in providing appropriate solutions for the management of this new format of records". Other earlier discussions such as the one held in Vienna, Austria on 26 August 2004 between some members of Africa Branch of the International Council on Archives International Records Management Trust (IRMT) and the National Archives of England and Wales have also emphasised the need for effective records management in Africa with respect to capacity building in the area of e-records management. In the same vein, earlier e-readiness assessments undertaken by SADC E-readiness Task Force in 2002 also underscored similar challenges. These include staff competencies, skills and tools needed to manage e-Business processes (Wamukoya & Mutula, 2005).

The literature also revealed that adopting integrated electronic information systems in government and organisation's transactions, e-records management policy formulation and implementation, establishing more training outlets for records managers and archivists, developing metadata for locating records which will go a long way incorporating ICT's infrastructure in managing e-records.

There is lack of literature on e-records readiness in the ESARBICA region as few studies have been conducted. The dearth of literature on e-records readiness is evidenced by the observation that most empirical studies in Africa and Eswatini in particular, have tended to focus more on paper-based records management.

In Eswatini particularly, the need for e-records readiness assessments in the context of e-government is necessitated by the fact that the 2013-2017 Eswatini e-government strategy does not consider the relationship between e-records readiness and e-government. With e-government being implemented in Eswatini government ministries, e-records that document the ministries transactions and online activities are also being produced. It is therefore important that the ministries ensure that standards are developed and implemented; that appropriate facilities are created and that adequate resources are invested in managing official records in electronic formats. The study assessed in detail the essence of factoring in e-records readiness when implementing the national e-government strategy. This would assist in sensitizing of policy makers especially those in government ministries and the cabinet on the essence of assessing and ascertaining e-records readiness for purposes of harmonizing them with the national e-government strategy. The study also revealed that e-government strategy should not exist in isolation, but it should be supported by or embedded on a robust national ICT policy.

METHODOLOGY

The study used to a large extent a quantitative approach and employed a survey design. This was, however, complemented by methodological triangulation of both quantitative and qualitative data collection methods to assess e-records readiness in government ministries in Eswatini. Surveys are largely quantitative and have been a widely used method in records management research. Ambira (2016) utilised survey research in his study to investigate the development of a framework for management of e-records in support e-government in Kenya. Another study by Marutha (2016) also utilized survey research to investigate the development of a framework to embed medical records management into healthcare service delivery in the Limpopo province of South Africa. The study employed a survey design in order to "describe, compare, contrast, classify, analyse and interpret implications of the findings".

In this study, no sampling was done and all the 18 government ministries, Cabinet office, Deputy Prime minister office, Eswatini National Archives, Department of Computer Services and the office of e-government department. The participants were the action officers (action officers refers to people who are working in the different ministries administration), registry personnel, the Director of National Archives, the Director of Computer services and Director of e-government. In determining the sample sizes for registry staff, Israel formula for determining sample sizes was used (Israel, 1992).

$$n = \frac{N}{1 + N(e)^{\wedge 2}}$$

Where n = desired sample size

N = Population size

e = Margin of error

e = ± 10%

90% Confidence level

The values of e=±10% and 90% confidence level were adopted. Consequently, using the Israel formula, the following samples were generated for records/registry staff:

Sample for Registry Staff

n= N

1+N (e) ^{2}

n=498

1+498 $(0.10)^2$

= 83 registry officers

The Action Officers like their records/registry counterparts were stratified random selected taking care to include all three management levels followed by random selection within each management level using the Ministries' organizational structure as the sampling frame. This resulted in 126 Action Officers. The distribution of the Action Officers was as follows: 42 top level management, 42 middle level management, and 42 from lower level management. The Ministries and sampled staff that were included in the study are reflected in Table 1

FINDINGS AND DISCUSSIONS

The findings and discussions that follow focus on the national legal and policy framework; electronic records management products and technologies; internal awareness of link of e-record management with e-government strategy, and the status of e-government in the government ministries.

National Legal and Policy Framework
Governing Management of e-Records

The study sought to find the national legal and policy framework governing management of electronic records in government ministries in Eswatini. These consist of statutes, laws, regulations, codes of conduct, best practice guidelines and ethics governing the business environment that relate to records management.

The study revealed that 123 (75%) of the respondents are unaware of Eswatini's National Archives Act no.5 of 1971 as a regulatory tool for records in the different ministries, while 40 (25%) are aware of the Act which is includes of records officers as depicted in Figure1.

Table 1. Government ministries and staff included in the study

Unit	Records/registry staff	Action Officers	Directors
Eswatini National Archives			1
Department of Computer Services			1
Department of E- Government			1
Cabinet office	4	6	
Deputy Prime minister office	4	6	
Ministry of Justice	4	6	
Ministry of Labor and Social Welfare	4	6	
Ministry of Public Service	4	6	
Ministry of Tourism	4	6	
Ministry of Works & Transport	4	6	
Ministry Natural Resources	4	6	
Ministry of Education	4	6	
Ministry of Finance	4	6	
Ministry of Housing and Urban Development	4	6	
Ministry of Home Affairs	4	6	
Ministry of Sports, Culture and Youth Affairs	4	6	
Ministry of Health	4	6	
Ministry of Information, Communications and Technology	4	6	
Ministry of Economic Planning	4	6	
Ministry of Commerce	4	6	
Ministry of Tinkhundla	4	6	
Ministry of Foreign Affairs	4	6	
Ministry of Agriculture	4	6	
Total	**n= 83**	**n=126**	**n= 3**

The study also sought to find out if the respondents 40 (25%) were aware of the contents of the legislation for records management. Figure 2 shows that 26(65%) were aware of the contents of the legislation and 14(35%) were not aware of the contents of the legislation.

Although 40 (25%) of the respondents were aware of the national legislation, it is quite concerning to note that there exists no national records management Act to guide the effective management of e-records. This is despite the fact that the

Figure 1. National legal and policy framework governing management of e-records

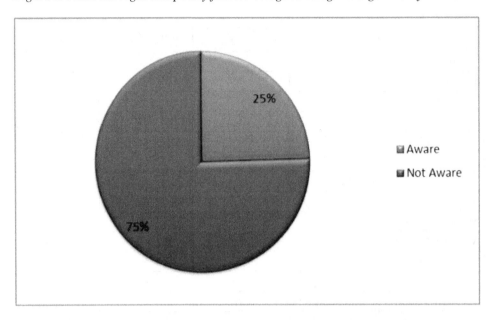

Figure 2. Awareness of the contents of the legislation for records management

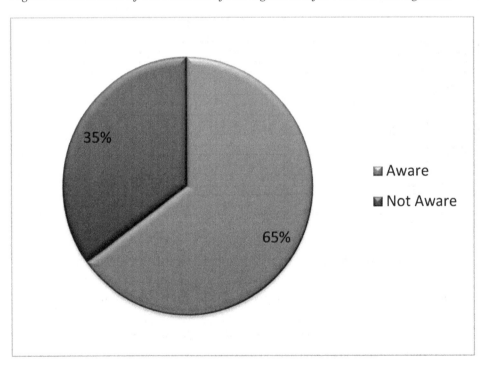

National Archives Act No.5 of 1971 focuses more on the archival stages of the Records Lifecycle, (Eswatini Government, 1971). The study also reveals that there is a National Archives and Records Management Bill of 2010, which captures the total Life Cycle management of all records regardless of media and format which has to be passed into law. The glaring lack of suitable legislative framework, the creation, maintenance, and long-term preservation of and access to e-records is left to chance.

Compliance to Policies

The study sought to find out whether the ministries had policies to guide the management of e-records. As shown in Figure 3 a total of 88 (53%) respondents acknowledged the non-existence of policies for managing e-records while 23 (14%) respondents acknowledged the existence of policies but did not know the major areas the policy covered. Some 52 (32%) respondents were not sure whether a policy for managing e-records existed.

Figure 3. Policies for managing e-records

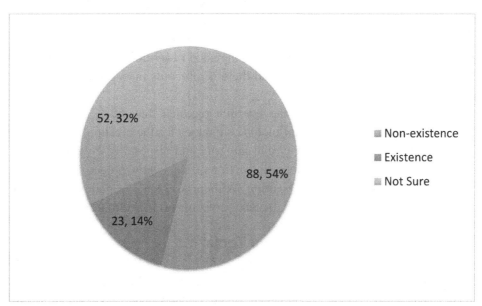

The study confirmed that a policy for managing e-records was existent. Interviews with the Director for Eswatini National Archives revealed that ENA has developed and distributed a national records management policy guideline, to help the ministries to develop internal records management policies throughout government ministries.

"There is a National Records Management Policy that has been developed by ENA and circulated to all government ministries and departments so that they can develop their own policies".

ISO 15489-1 (2016) (section 5) stipulates that a records management policy and procedures of an organisation should demonstrate the application of the regulatory environment to their business processes. Section 6 specifies that an organisation should "establish, document, maintain and promulgate policies, procedures" to guarantee that "its business need for evidence and accountability and information about activities is met". The study therefore confirms that the guiding national records management policy has not been used by government ministries. The study also established that the Ministry of ICT had developed an ICT Policy and e-government strategy, however it did not address electronic record keeping issues. The e-government strategy emphasises that e-government is a vehicle for national economic and social development by ensuring effectiveness, efficiency, transparency and accountability on the part of the government, but it does not highlight whether e-records ready in government ministries for purposes of use in the implementation of e-government. On the other hand, the ICT policy addresses issues such as, the ICT infrastructure policy, policy compliance and sustainability and procurement, maintenance and disposal of ICT infrastructure and systems.

The Director of Computer Services explained that "there is an ICT policy in place and an e- Government strategy but they don't capture the creation, receipt, use and disposal of records but instead there is a records management procedure which captures that".

Through document analysis, the research revealed that the ICT policy and the e-government strategy did not include strategies for the creation, receipt, use and maintenance, storage, security and integrity and disposal of e-records. Such strategies will guide records officers and action officers in the proper management of e-records from creation to disposition. Without a strategy or policy in place, it becomes difficult for the ministries to manage records in an electronic environment.

Compliance to Procedures and Tools

The study sought to ascertain the level of compliance to procedures and tools. The study revealed that there are documented records management procedures manual developed by Eswatini National Archives which are used by the ministries. However, the study revealed that the procedures manual caters for physical records and it is not clear for e-records. Nevertheless, the study has revealed that 18 (25.7%) records officers that comply to the procedures manual indicated that the manual was 'Above average' while 29 (41.4%) said it was below average. This is shown below in Table 2.

Table 2. Compliance to records management procedure and tools [N=70] (Records officers)

Response	Frequency	Percent
Above average	18	25.7
Average	23	32.8
Below Average	29	41.4
Total	**70**	**100**

Electronic Records Management Products and Technologies

The study sought to find out whether technology for electronic records management was available in the ministries. Respondents were presented with a list of electronic technologies and asked to tick against the ones that were available in their offices. Table 3 shows that 111 (69.3%) indicated that they have computers in their offices; 140 (87.5%) indicated that they have mobile phones; 10 (6.25%) digital camera and 2 (1.25%) EDRMS.

Table 3. Technologies for electronic records management (N160)

Response	Frequency	Percentage
Computers	111	69.3
Printer, scanners, photocopiers, laminators	70	43.75
CD, CD-ROM, DVD, VCD, Flash Drive	109	68
Electronic document records management system (EDRMS)	2	1.25
Internet connectivity	98	61.25
Online transactional processing systems (OLTPS)	35	21.9
Mobile phones	140	87.5
Decision support systems (DSS)	0	0
Digital camera	10	6.25
Cassette recorder and tapes	35	21.9

*Multiple responses were possible

Strategies Used to Create and Receive e-Records

The respondents were also asked to indicate strategies they use to create and receive e-records in their officers. The question was directed to registry staff and action officers whose responsibilities include the creation and receipt, use, maintenance and disposal of e-records as part of their day to day business activity in the ministries. Table 4 shows that 151 (94%) respondents made printed copies of the e-records they created while 151 (94%) made printed copies of the official records they received. One hundred and eleven (69.3%) create and save on computer files while 51(32%) receive and save e-records on the computer hard disk. One hundred and nine (68%) of the respondents create and save on storage devices such as CD and USB and while 12 (9%) receive and store e-records on storage devices.

Table 4. Strategies used to create and receive e-records[N=160]

Response	Frequency	Percentage
Create and save on computer files	111	69.3
Create and save on storage devices such as CD and USB	109	68
Receive and save on the computer hard disk	51	32
Make printed copies after receipt	151	94
Create and make printed copies	151	94
Receive and store on storage devices	12	9

*Multiple responses were possible

The findings indicate that there is no standardised procedure put in place for the effective management of e-records across Eswatini Government Ministries. This can be attributed to the general practice that most e-records (including e-mails) were created and then filed as paper-based records. Such a situation is not good especially if the e-records will exist as corporate memory of those ministries. The study also observed that each office that created electronic records had its own way of maintaining, retrieving, and storing electronic records. In some offices, memory sticks were found lying on top of tables without protective lids to minimize their exposure to dust.

The researchers also observed that the majority of respondents made printed copies of the records they created and received. This could be attributed to the fact that despite computerization of some of the ministries and departments, those ministries had not done away with the use of paper records as a means of transacting business.

Respondents maintained soft copies of the e-records they created and received. The strategies used to create and receive official e-records by the ministries were individual measures that were undertaken by the respondents without necessarily involving their ministries or Eswatini National Archives as a watchdog over the creation, maintenance, use, and disposal of records. It was apparent therefore, that the creation and receipt of e-records did not adhere to any records management principles or policy.

A follow up interview with the Director of ENA and Director Computer Services revealed that ENA is piloting an Electronic Document Records Management (EDRM) solution to the Ministry of ICT and the Cabinet office to effectively manage and preserve government records as corporate memory for future generations. The main aim is to bring uniformity and standardization of e-records systems and management practice across government ministries and departments. The study observed that although there is a system that is being piloted, some respondents kept printed copies of e-records in desk drawers and cabinets without necessarily filing the records. Paper records had continued to clog the office space thus, resulting in the in-accessibility of records whenever they were required for reference.

Strategies Used to Access e-Records

The study sought to establish how officers accessed information contained in e-records. The results revealed that 151 (94%) respondents made printed copies of e-records and filed copies manually in folders to facilitate access while 109 (68%) respondents used storage devices such as USB sticks and CDs as a strategy to ensure that whenever the information was required it was made available in the Ministries and Departments. 60 (40.6%) of the respondents indicated that they used backup while 98 (61.25%) respondents used electronic mail to distribute e-records. The results are presented in Table 5.

Table 5. Strategies used to access e-records(N=160)

Response	Frequency	Percentage
Making printed copies	151	94
Storage devices such as USB sticks and CDs	109	68
Back up	65	40.6
Electronically via mail	98	61.25

*Multiple responses were possible

The researchers also established that some respondents use personal folders to store e-records. Such respondents did so as a personal initiative and gave the folders names that were only known to them. The study also noted that there were no procedures in place to provide guidance on the management of computer files. Without the assistance of records creators or the persons who received the e-record, it is impossible to access or retrieve the information. At times officers were having difficulties retrieving the e-records they stored on computer folders because they had forgotten the file name(s) and the location of the folder(s).

Status of e-Government in the Government Ministries

The study also investigated the status of e-government in the ministries and it established that the ministries were at the initial phase as regards e-government implementation. The study established that Eswatini government ministries were just starting to be prepared for the implementation of e-government services by way of putting in place the necessary infrastructure and operating administrative functions of the ministries electronically.

The majority of the action officers keep their records in their offices whose existence no one else knows about. No procedures are followed when action officers file documents which include e- records where the use of folders and naming conventions is not systematic. The e-recordswere neither well-arranged nor well documented, causing problems when action officers wanted to retrieve records. The findings indicate that there is no standardised procedure put in place for the effective management of e-records across Eswatini government ministries. This can be attributed to the general practice that most e-records (including e-mails) were created and then filed as paper-based records. The study also observed that each office that created e-records had its own way of maintaining, retrieving, and storing the e-records.

CONCLUSION

The study established that the level of e-records readiness in the government ministries is at an infant stage. E-records management is disjointed, haphazard and poorly handled. Staff display poor records management skills and there is lack of professional training of staff, a weak legislation and policy framework, absence of a disaster preparedness plan, slow progress in the implementation of EDRMS and low capacity building as records management staff is rarely taken for training. There is also inadequate senior management support. The study has also revealed that opportunities for increasing the depth of e-records readiness exist such as:

availability of financial resources for EDRMS project. In view of these, the study makes the recommendations below on how management of e-records could be improved in the government ministries in Eswatini.

- There is an urgent need to fast track the amendment and passing into law of the proposed National Archives and Records Management Bill of 2010 by the Director of Eswatini National Archives, which captures the total Life Cycle management of all records regardless of media and format.
- National Records Management Policy be formulated by Eswatini National Archives to regulate and streamline the effective management of e-records so that they can survive as corporate memory of Eswatini government transactions just as paper-based records have been treated all along.
- ENA should facilitate the development of records management policies for the ministries which should include the management of e-records in e-government, requirements for systems for managing electronic records, preservation of electronic records, common data to be shared across the ministries, ICT infrastructure requirements for managing electronic records and email management.
- ENA, in collaboration with the Department of Computer Services and the Department of e-government should develop specifications and functional requirements for e-records management systems within the context of e-government in Eswatini to ensure that all software acquired for managing e-records capture the requirements of e-government, including capability to interface with the e-government platforms for push and pull of data.
- Records management procedures have to be developed to enable the smooth operation of the ministries registries which has been used as a dumping site for action officers. Since the working space is not adequate in some offices, records have to be kept in the ministries registries where they can be retrieved when required by everyone. These would also allow smooth of information sharing and there will be proper documentation of all files that exist in the registries. The retrieval of files in the registries would be much easier and duplication of records would be reduced.
- ENA should regularly conduct sensitization seminars and workshops for raising awareness in the ministries for the different categories ranging from senior management, middle management, lower management and the rest of the staff.
- ENA should regularly conduct sensitization seminars and workshops for raising awareness in the ministries for the different categories ranging from senior management, middle management, lower management and the rest of the staff.

- ● ENA should take a lead role in developing a comprehensive preservation strategy that harmonizes management of electronic records across government ministries.

REFERENCES

Ambira, C. (2016). *A Framework for management of electronic records in support of e-Government in Kenya* (PhD Thesis). Pretoria: University of South Africa.

Common Wealth Secretariat. (2013). *e-government Strategy for the Eswatini: 2013 to 2017*. Available from: http://www.gov.sz/images/e-government%20 strategy%20 final%20document%20that%20was%20adopted.pdf

Eswatini, Government. (2015). *Department of Eswatini National Archives. Report on the Review of government registries in Eswatini*. Mbabane: Government Printer.

European Commission. (2001). *Model requirements for the management of electronic records. Bruxelles-Luxemburg*. European Commission.

Ginindza, M. (2008). *The state of e-government in Eswatini with special reference to government ministries and departments* (MA Thesis). University of Kwazulu Natal.

International Records Management Trust (IRMT). (2003). *The World Bank Development Facility: Evidence-Based Governance in the Electronic Age: Call Study Summaries: Personnel and Payroll Records and Information Systems in Tanzania, 2002*. Available from http:www.irmt.org/download/DOCUM%7E1/GLOBAL/ Case%20Studies%Summaries.pdf

International Records Management Trust (IRMT). (2004). *The e-recordsreadiness tool*. Available from http://www.nationalarchives.gov.uk/rmcas/downloads.asp# additional tools

International Records Management Trust (IRMT). (2005). *IRMT e-readiness assessment model*. London: IRMT. Also available: www.irmt.org

International Records Management Trust (IRMT). (2009). *Training in Electronic Records Management: Understanding the Context of Electronic Records Management*. Available from: http://www.irmt.org/educationTrainMaterials

International Standards Organisation (ISO). (2016). *ISO 15489-1, Information and Documentation – Records Management Part 1: General*. Geneva: International Standards Organisation.

Israel, G. D. (1992). *Determining sample size*. Fact Sheet PEOD-8. Available from http://www.edis.ifas.ufl.edu

Kalusopa, T. (2011). *Developing an e-recordsreadiness framework for labour organisations in Botswana* (PhD thesis). Pretoria: University of South Africa.

Kamatula, G. A. (2010). E-government and e-records: Challenges and prospects for African records managers and archivists. *ESARBICA Journal, 29*, 147–163.

Keakopa, M. (2009). A critical Review of the Literature of Electronic Records Management in the ESARBICA Region. *ESARBICA Journal, 28*(1), 78–104. doi:10.4314/esarjo.v28i1.44398

Kennedy, J., & Schauder, C. (1998). *Records management: a guide to corporate record keeping* (2nd ed.). Melbourne: Longman.

Kulcu, O. (2009). Evolution of electronic records management practices in e-government: A Turkish perspective. *The Electronic Library, 27*(6), 999–1009. doi:10.1108/02640470911004084

Marutha, N. S. (2016). *A framework to embed medical records management into the healthcare service delivery in Limpopo province of South Africa* (PhD thesis). Pretoria: University of South Africa.

Maseko, A. (2010). *Investigated the management of Audiovisual Records at the Eswatini Television Authority (STVA)* (M.A. thesis). Gaborone: University of Botswana.

Mnjama, N., & Wamukoya, J. (2004). E-governance: the Need for an E-records Readiness Assessment Tool. *Proceedings of the SADC Regional Consultation on National e-government Readiness*. Available from http://www.comnet-it.org/news/CESPAM-Botswana.pdf

Mnjama, N., & Wamukoya, J. (2007). E-government and Records Management: An Assessment Tool for E-recordsReadiness in Government. *The Electronic Library, 25*(3), 274–284. doi:10.1108/02640470710754797

Moloi, J. (2006). *An investigation of e-recordsmanagement in government: case study of Botswana* (MA thesis). Gaborone: University of Botswana.

Moloi, J., & Mutula, S. (2007). E-records management in an e-government setting in Botswana. *Information Development, 23*(4). Available from http://idv.sagepub.com/content/23/4/290

NARA. (2006). *Managing electronic records in governmental bodies: policy, principles and requirements.* Republic of South Africa: Department of Arts and culture.

Nengomasha, C. (2009). *A study of electronic records management in the Namibian public service in the context of e-government* (PhD thesis). Windhoek: University of Namibia.

Nolan, B. (2001). *Conclusion: Themes and Future Directions for Public Sector Reforms in Public Sector Reform: An International Perspective.* New York: Palgrave Macmillan.

Tsabedze, V. (2011). *Records Management in government ministries in Eswatini* (MA thesis). University of Zululand, Kwa-Ndlengezwa.

United Nations. (2001). *Benchmarking e-government: a global perspective assessing the progress of the UN member states.* http://www.unpan.org/egovernment.asp

Wamukoya, J., & Mutula, S. (2005). Capacity-building requirements for e-records management: The case in east and southern Africa. *Records Management Journal, 15*(2), 71–79. doi:10.1108/09565690510614210

White House. (2012). *Benefits of e-government initiatives. Annual report to the congress.* Office of Management and Budget. https://www.whitehouse.gov/sites/default/files /omb/assets/egov_docs

Xiaomi, A. (2009, July). An integrated approach to records management. *Information Management Journal,* 24-30.

Chapter 5
Electronic Participation in a Comparative Perspective:
Institutional Determinants of Performance

Antonio F. Tavares
ⓘ https://orcid.org/0000-0003-4888-5285
*School of Economics and Management, University of Minho, Portugal &
Operating Unit on Policy-Driven Electronic Governance, United Nations
University, Portugal*

João Martins
*NIPE, University of Minho, Portugal & Operating Unit on Policy-Driven
Electronic Governance, United Nations University, Portugal*

Mariana Lameiras
ⓘ https://orcid.org/0000-0002-9134-9296
*Operating Unit on Policy-Driven Electronic Governance, United Nations
University, Portugal & Communication and Society Research Centre, University
of Minho, Portugal*

ABSTRACT

Electronic participation can play a crucial role in building broader public involvement in decision-making and public policy to bring about more inclusive societies. Prior empirical analyses have neglected the fact that political institutions are not only affecting the expansion of digital government, but also often interact with more structural conditions to constrain or incentivize the adoption and expansion of e-participation. This research analyses the role of institutional factors in encouraging or constraining e-participation across countries. Fractional regression models are

DOI: 10.4018/978-1-7998-1526-6.ch005

employed to analyze panel data (2008-2018) from the United Nations Member States scores in the E-Participation Index (EPI) developed by the United Nations Department of Economic and Social Affairs (UNDESA). The results indicate that the quality of democratic institutions, freedom of the press, and government effectiveness are all relevant predictors of a higher performance in e-participation. Policy implications are drawn in line with the 2030 UN Agenda for Sustainable Development Goals.

INTRODUCTION

Electronic participation (e-participation) has the potential to facilitate citizens' involvement in public affairs, whether through information provision, expanded consultation or in-depth deliberative decision-making processes. An early definition of e-participation describes it as "ICT-supported participation in processes involved in government and governance" (OECD, 2003). Saebø et al. (2008, p.402) define e-participation to include all forms of "technology-mediated interaction between the civil society sphere and the formal political sphere and between civil society sphere and the administration sphere". Highlighting the massive growth in academic research and governmental practice of e-participation, Wirtz, Daiser, & Binkowska (2018, p. 3) define it as "a participatory process that is enabled by modern information and communication technologies" and involves "stakeholders in the public decision-making processes through active information exchange, and thus fosters fair and representative policy-making".

Often portrayed as a field lacking consistency, e-participation has become increasingly popular over the past few years (Medaglia, 2012). In fact, with globalization and technological innovations, participatory processes are being challenged and the evolving technology requires stakeholders to continuously 'chase the digital wave' (Gibson, Römmele, & Williamson, 2014) and to foster ways of promoting 'creative citizenship' (Rodríguez Bolívar, 2018). The participation literature highlights individual resources and the role of institutional and political factors as determinants of participation (Verba, Schlozman, & Brady, 1995). More recently, revisited versions of this theory encompass the role of digital technologies and the positive correlation between individual resources and the likelihood of online engagement (Anduiza, Gallego, & Cantijoch, 2010). More generally, the academic literature in the field reports positive effects of e-participation for democracy, inclusion, transparency, accountability and good governance (Bertot, Jaeger, & Grimes, 2012; Medaglia, 2012; Noveck, 2009; Wirtz et al., 2018).

Following the United Nations E-Government Survey, it is possible to identify three dimensions of e-participation, namely e-information, e-consultation and e-decision-making (United Nations, 2018). E-information reflects government uses of digital technology to provide information to citizens. Information made available through Information and Communication Technologies (ICTs) can then be used as evidence for the advancement of the next stages in e-participation: consultation and decision-making. Public policies and the provision of services can incorporate the suggestions and commentaries of citizens directly or indirectly affected. When government elicits citizen participation in the formation of public policies and service delivery choices using ICTs, the process is defined as e-consultation. Once the consultation period is over, public officials "analyze the comments received and publish overall findings" (Scott, 2006, p. 350). The third stage of e-participation involves citizen participation in decision-making employing ICTs, including e-voting, online deliberation systems, and the evaluation of public policy proposals using social media (United Nations, 2018).

Early research on the determinants of the progress of e-government and e-participation in countries around the world highlights the role of technical infrastructure, economic development, and education levels as prime explanations (Åström, Karlsson, Linde, & Pirannejad, 2012; Siau & Long, 2009). However, prior empirical analyses have largely failed to take into account the institutional framework under which these progresses have been accomplished (Gulati, Williams, & Yates, 2014). More importantly, these analyses have neglected the fact that political institutions are not only affecting the expansion of digital government, but also often interact with more structural conditions to constrain or incentivize the adoption and expansion of e-participation (Gulati, Williams, & Yates, 2014; Kneuer & Harnisch, 2016).

This work contributes to the literature in several ways. First, it investigates how institutional differences – autocracies vs. democracies, public trust in elected officials, absence of corruption, freedom of the press, and government effectiveness – account for the variation in e-participation scores across countries and over time. This builds on and extends prior work by Bussell (2011) and Kneuer & Harnisch (2016) assessing the differences in the expansion of e-government in democracies and autocracies. Second, this aggregate analysis allows the examination of not only the direct effects of institutional variation on e-participation, but also the interactions between institutional conditions and other factors affecting the development of e-participation across countries. Third, by focusing the analysis on the political and institutional determinants of e-participation at the country level, this chapter fills in an important lacuna in prior empirical studies, which have focused primarily on socioeconomic resources and internet skills as individual drivers of e-participation (Khoirunnida, Hidayanto, Purwandari, Kartika, & Kosandi, 2017; Vicente & Novo, 2014). Lastly, our contribution is also methodological. The analyses employ fractional regression models (Papke & Wooldridge, 1996) as the main empirical method to avoid the pitfalls entailed in using ordinary least square regression on a censored dependent variable like the UN E-Participation Index.

In particular, this research analyses the effects of institutional factors in encouraging or constraining e-participation across countries over a period of ten years (2008-2018) through a quantitative approach. The dependent variable of the analysis is a proxy for e-participation readiness, the UNDESA's E-Participation Index (EPI). This index evaluates 193 countries every two years, based on three main dimensions: provision of information by governments to citizens (e-information), interaction with stakeholders (e-consultation), and engagement in decision-making processes (e-decision making). The key explanatory variables of this research are the institutional and political factors affecting e-participation.

After the introduction, the theoretical model is presented, and the hypotheses supported by the literature on e-participation are discussed. The following section introduces the data and methods employed in this research and then the empirical analyses are presented. The next part of this chapter is devoted to a discussion of the findings. Lastly, the authors focus on a set of conclusions and directions for future research.

THEORETICAL MODEL

This chapter builds on prior studies of factors affecting e-government development and proposes a theoretical model to examine how the performance of political institutions impacts the adoption and implementation of e-participation in a comparative perspective. The model highlights not only the direct effects of political institutions, but also how their performance interacts with technology penetration and socio-economic development to account for specific levels of e-participation both across space and over time. Concretely, the model posits that both variables mediate the positive effects of the quality of political institutions on e-participation.

Early work in the field of Information Systems has identified a series of factors that impact e-government development in countries around the world. Several studies have shown that technology penetration and human development levels are positively associated with a country's e-government development (Siau & Long, 2009), performance (Stier, 2015), and maturity (Ifinedo, 2012; Ifinedo & Singh, 2011; Larosiliere & Carter, 2016; Singh, Das, & Joseph, 2007). These efforts have proven useful for the theoretical developments attempting to explain the expansion of e-participation. One of the most comprehensive pieces of research by Krishnan et al. (2017) combines the analysis of the determinants of e-government maturity and e-participation in a single article and confirms the positive association of both ICT infrastructure and human capital development with the expansion of digital government. If a country's human and socio-economic development is an indicator of a stronger civil society and more active grassroots movements, it is also likely that these countries will display higher commitment to e-participation (Reddick and Norris, 2013).

Early work by Milner (2006, p. 178) suggests that "political institutions in particular matter for the adoption of new technologies because they affect the manner and degree to which winners and losers from the technology can translate their preferences into influence". Democracies are also more likely to experience citizen pressure and transparency norms capable of stimulating the use of e-participation tools (Kneuer & Harnisch, 2016). In a similar vein, Stier (2015) suggests that liberal democracies with competitive elections are more likely to be concerned with citizen-centric deliberation and therefore promote e-participation. Democratic governments are also more likely to encourage multiple forms of political expression and checks on power (Gulati, Williams, & Yates, 2014). This willingness to promote transparency will find in e-participation tools the obvious means to attain these goals.

Milner (2006, p. 178) also argues that "the Internet can provide civil society with uncensored information, costless sharing of that information, and tools to overcome collective action problems for organizing opposition". Logically, other indicators of the quality of political institutions are just as likely to be related to e-participation. Just as e-government in general, e-participation has also been associated with perceived transparency (Zheng & Schachter, 2017), reduced corruption (Bussell, 2011), improved trust in government (Zolotov, Oliveira, & Casteleyn, 2018), and enhanced legitimacy of political systems through the involvement of citizens in the political and administrative debate (Åström et al., 2012; Stier, 2015).

All these indicators of the quality of political institutions have been associated with e-participation, but their interaction with more infrastructural conditions is less understood. In other words, the quality of political institutions is likely to have varying effects on e-participation across countries with different levels of technology penetration and socio-economic development. Authors investigating the determinants of e-participation have failed to explore this possibility, since they have only tested the direct (or additive) effects of these variables. However, it is reasonable to expect that countries with similar political institutions or comparable democratic performance indicators will perform very differently if the penetration of technology or human capital levels are substantially different. In fact, high quality political institutions are more likely to rely on e-participation tools if their infrastructure and human capital performance is also high. Conversely, autocracies with large investments in technological infrastructure and penetration may fare better in terms of e-participation than autocracies where this investment has not occurred. In theory, both technology penetration and socio-economic development are likely to mediate the relationship between the quality of political institutions

Figure 1. Theoretical model of E-participation

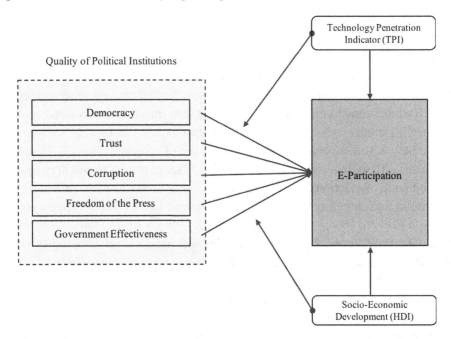

and the reliance on e-participation. Figure 1 displays these theoretical relationships. The next part of this chapter operationalizes the concepts included in the model and translates the theoretical connections into hypotheses.

HYPOTHESES

The model posits a positive association between higher quality of political institutions and the levels of e-participation. Research exploring these effects on e-participation has employed a diverse set of indicators to operationalize the quality of political institutions (Stier, 2015), including democratic performance, corruption levels, trust in government, government effectiveness, and freedom of the press, and obtained mixed findings. This part of the chapter reviews this literature and employs it to support the hypotheses to be tested in the empirical section. The first part of this section discusses the additive hypotheses linking the quality of political institutions and e-participation. In the second part, we present the interactive hypotheses arguing that the relationship between political institutions and e-participation is mediated by the countries' structural conditions, namely varying levels of technology penetration and socioeconomic development.

Additive Hypotheses

Aggregate analyses of the effects of democratic performance at the country level report interesting but somewhat limited findings. Kneuer & Harnisch (2016) found substantive differences in e-participation levels between democracies and autocracies. Democracies are early adopters of e-participation and remain above all other regime categories defined by the authors, including flawed democracies, multiparty, single party and military regimes. Åström *et al.* (2012) employ ordinary least squares regressions to analyze the effect of a country's democratic performance on e-participation levels over time. They find a positive association between their measure of democracy based on the Freedom House and Polity IV Indexes and the 2003 UN E-Participation Index. This effect disappears in the remaining years of the analysis (2004, 2005, 2008 and 2010). A similar finding is reported by Stier (2015), who finds positive associations between the level of democracy and e-government performance for the years of 2002, 2003, 2004 and 2007 and null findings after that (2009, 2011, and 2013). This dynamic analysis also shows that autocratic governments may be the ones most interested in improving the interaction with citizens using e-participation tools. Gulati et al. (2014) use multiple regression analysis to explain e-participation capabilities across countries and fail to find the expect positive effect of a democratic political structure. A study by Jho & Song (2015) uncovers a positive association between the level of democracy and e-participation, but the same authors fail to confirm a similar effect for the level of institutionalization of free speech and association. Despite these mixed findings, the hypothesis regarding democratic performance reflects the theoretical expectations stated above (Stier, 2015):

H1a: More consolidated democracies display higher levels of e-participation.

In contrast with the aggregate analyses mentioned above, individual-level analyses report important links between political variables and individual e-participation. Porumbescu (2016) employs a sample of 1100 Seoul residents and finds a positive relationship between citizen perceptions of public sector trustworthiness and the use of public sector social media accounts. Zolotov et al. (2018) conducted a meta-analytical review of e-participation adoption models and found that generalized trust, and more specifically, trust in government are significant predictors of the likelihood of adoption of e-participation. Novo Vázquez & Rosalía Vicente (2019) analyze e-participation in Spanish municipalities and find that political interest, external political efficacy, and associational membership are relevant predictors of individual e-participation. We extend these tests to the aggregate level by hypothesizing that:

H1b: Higher levels of trust in politicians have a positive effect on a country's level of e-participation

Bussell (2011) investigated the association between the corruption scores measured by Transparency International's Corruption Perceptions Index and a country's e-government quality as assessed by the UN E-government Index. The author finds a robust relationship between both variables, therefore supporting the idea that corruption levels dampen countries efforts to promote "higher quality technology-enabled service reforms" (p.275). Given the limited evidence linking corruption levels and e-participation, we might expect a similar relationship:

H1c: Higher corruption levels have a negative effect on a country's level of e-participation.

To our knowledge, no empirical study has yet linked freedom of the press levels with e-participation. However, early work by Sylvester and McGlynn (2010) analyzed data from the 2007 Pew Internet and American Life project and found an association between using the newspaper for information and the likelihood of contacting government by email. Hollyer, Rosendorff and Vreeland (2014) developed the HRV government transparency index and applied it to 149 countries from 1980-2008. The authors test the effect of daily newspaper circulation on the HRV index using World Bank data and found support for a strong positive effect. While both articles are not testing the relationship between freedom of the press and e-participation, they provide anecdotal evidence supporting our next hypothesis:

H1d: Freedom of the press is positively associated with a country's level of e-participation.

In contrast with the mixed findings reported for the association between democratic performance and e-participation levels, higher government effectiveness is systematically associated with e-participation (Gulati et al., 2014; Stier, 2015). In fact, Stier's analyses indicate an increasing impact of this explanatory factor over time. Gulati, Williams, & Yates (2014) investigate the determinants of e-participation in 158 countries reported by the 2010 UN E-Participation Index. The authors find that countries with a more effective public sector governance display higher scores of e-participation. This result underscores how the professionalization of public administration helps governments to embrace novel online participatory tools. Conversely, it also suggests that weak government institutions compromise the best intentions to undertake innovative e-participation opportunities. This suggests that:

H1e: Better government performance (effectiveness) is positively associated with a country's level of e-participation.

The diffusion of e-government (and e-participation) can be hindered by restricted access to broadband internet bandwidth, mobile network coverage, and technological interoperability (Zhang, Xu, & Xiao, 2014). Sound and reliable ICT infrastructure is even more crucial for the implementation of e-participation tools, since it facilitates access to information, reduces physical and geographical barriers to participation, improves the quality of feedback in public consultations, and empowers citizens to engage in deliberative policy-making (DiMaggio, Hargittai, Neuman, & Robinson, 2001; Saebø et al., 2008). Several empirical studies have investigated the impact of technology infrastructure on e-government development and maturity levels (Ifinedo & Singh, 2011; Siau & Long, 2009; Singh et al., 2007; Stier, 2015). Åström et al. (2012) investigate the determinants of e-participation in over 100 countries between 2003 and 2010 and find that internet users per 100 citizens – a proxy for access to technology infrastructure – is the strongest predictor of e-participation. Krishnan, Teo & Lymm (2017) employ cross-sectional data from 183 countries and find a positive effect of ICT infrastructure on both e-government maturity levels and a government's willingness to implement e-participation. Given the strength of these findings, we predict that:

H2: The level of technology penetration in a country is positively associated with its e-participation levels.

Developed in the field of Economics, human capital theory argues that investments in education, training, knowledge and health of the individuals in the labor force lead to economic growth over time (Becker, 1964; Schultz, 1961). E-government development has been linked to these indicators of socioeconomic development in past empirical works (Siau & Long, 2009; Stier, 2015). However, prior attempts to test the relationship between socioeconomic development levels and e-participation have failed to provide consistent results. The study by Åström et al. (2012) finds no association between the Human Development Index (HDI) and the UN's E-Participation Index. Jho & Song (2015) find a positive correlation between the same variables for the 2010 version of the e-participation index, but their models are severely misspecified, so this result is unconvincing. Lastly, Gulati et al. (2014) examine several indicators of sociodemographic development (education, urbanization and land area) and find positive associations between these indicators and the e-participation index. The overall set of findings suggests that socioeconomic development is likely to predict the levels of e-participation, but this association seems largely contingent on the indicators employed to measure socioeconomic development. Hence, we hypothesize that:

H3: The level of socioeconomic development in a country is positively associated with its e-participation level.

The following paragraphs focus on the moderating effects that technology penetration and socio-economic development can have on the impact of the quality of political institutions.

Interactive Hypotheses

Prior attempts have been developed to explore the interaction between political institutions and other factors affecting e-participation levels. Jho & Song (2015) investigate the moderating effects between technology and institutions, but their analysis is based on a single point in time (data from the 2012 UN E-Participation Index). The authors find that technology reinforces the positive effect of political institutions on e-participation. Despite these earlier efforts, none of the empirical studies investigated multiple moderating effects between explanatory variables both across countries and over time.

The theoretical model portrayed in Figure 1 argues that technology penetration and the levels of human and socio-economic development of countries is likely to enhance the positive effects of the quality of political institutions on e-participation levels. In other words, it can be expected that better technology penetration and human capital will reinforce the relationship between political institutions and e-participation, whereas poor technology penetration and insufficient human capital are unsurmountable obstacles to any attempts at increasing e-participation levels. More concretely, we predict that:

H4: The positive association between the quality of political institutions and e-participation will be stronger in countries with better technology penetration.

And:

H5: The positive association between the quality of political institutions and e-participation will be stronger in countries with better human capital development.

Figure 2 displays the empirical model of e-participation to be tested in the remaining sections of this chapter.

Figure 2. Comprehensive research model for empirical studies

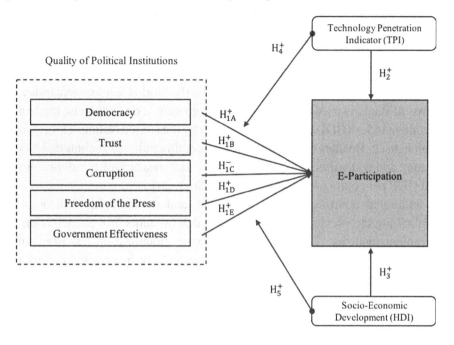

DATA AND METHODS

The dependent variable of the analysis is the UNDESA's E-Participation Index (EPI). The index is currently a biannual publication that evaluates 193 countries on three main dimensions: provision of information by governments to citizens (e-information), interaction with stakeholders (e-consultation), and engagement in decision-making processes (e-decision making). The collection of the data in which the EPI is based is performed by a group of more than 100 researchers, through a survey that evaluates the websites and portals of the central government, ministries and other governmental agencies of all UN member states.

In order to test the hypothesis related with relationship between institutional indicators and e-participation, the analysis employs five independent variables, assessing different dimensions of the political institutions, namely:

- The Autocracy-Democracy Index of the Polity IV database (*autocracy-democracy*);
- The Public Trust in Politicians Index by the World Economic Forum (*trust_politicians*);
- The Control of Corruption Index of the Worldwide Governance Indicators (*corruption*);

- The Freedom of the Press Index by Reporters Without Borders (*pressfree*);
- The Government Effectiveness Index of the Worldwide Governance Indicators (*gov_effectiveness*).

As a technology penetration indicator, the number of mobile cellular subscriptions per 100 inhabitants (*mobile_cellular*), taken from the World Development Indicators of the World Bank, is used. To proxy the socio-economic development, the Human Development Index (HDI), compiled every year by the United Nations Development Program is used. Besides the income dimension, this index also contemplates life expectancy and education, what makes it a broader measure of development than the real GDP per capita (Åström et al., 2012; Stier, 2015).

As additional control variables, economic and demographic indicators are used, following the previous literature on the determinants of e-participation and e-government maturity. The degree of economic globalization has been linked to increased investments in e-government, particularly among market-oriented autocracies, as suggested by Stier (2015). The rationale is two-fold. On one hand, autocracies hope to diffuse political criticism by delivering on the economic development front. Significant investments in e-government infrastructure and services are likely to accomplish this goal. On the other hand, authoritarian governments' search for legitimacy, both domestically and internationally, may stimulate the relaxation of information restraints through limited promotion of e-participation tools (Åström et al., 2012; Egorov, Guriev, & Sonin, 2009; King, Pan, & Roberts, 2013; Noesselt, 2014). Consequently, we use the KOF Index of economic globalization (*eco_global*) developed by Dreher, Gaston & Martens (2008) as independent variable. Lastly, the analyses also include the size of a country's population (*pop*) to capture economies of scale entailed by the fixed costs associated with the implementation of e-government programs (Bussell, 2011; Milner, 2006; Norris & Moon, 2005; Stier, 2015). Table 1 presents the descriptive statistics for all variables.

As the EPI is a censored variable, defined in a scale where 0 represents the lowest possible score and 1 the highest possible score, the analyses employ fractional regression models (Papke & Wooldridge, 1996) as the main empirical method. This allows us to overcome a pitfall of previous studies that use the least squares estimator (e.g. Åström et al., 2012; Jho & Song, 2015; Stier, 2015): the possibility of predicting outcome values for the dependent variable that are lower than 0 or higher than 1, violating the boundaries of the index. As a strategy to avoid multicollinearity and given the high correlation amongst some of the institutional variables, the H1 sub-hypotheses (H1a to H1e) are tested by estimating different models for each institutional dimension. Additionally, as the EPI is coded in the year prior to the index's release[1], as a strategy to prevent reverse causality and minimize endogeneity concerns, the

Table 1. Descriptive Statistics for the period 2008-2018

Variable		Mean	Standard Deviation	Observations
EPI	Overall	0.340	0.290	N=1145
	Between		0.220	n=191
	Within		0.190	
autocracy-democracy	Overall	0.699	0.312	N=1931
	Between		0.305	n=162
	Within		0.069	
trust_politicians	Overall	0.435	0.171	N=1642
	Between		0.157	n=149
	Within		0.061	
corruption	Overall	0.513	0.199	N=2265
	Between		0.197	n=189
	Within		0.030	
press_free	Overall	0.719	0.159	N=2017
	Between		0.144	n=172
	Within		0.067	
gov_effectiveness	Overall	0.487	0.197	N=2265
	Between		0.195	n=189
	Within		0.030	
mobile_cellular	Overall	89.131	42.656	N=2235
	Between		36.525	n=190
	Within		22.537	
HDI	Overall	0.689	0.156	N=2205
	Between		0.155	n=185
	Within		0.019	
pop (billions)	Overall	0.037	0.138	N=2274
	Between		0.138	n=190
	Within		0.005	
eco_global	Overall	0.585	0.155	N=2002
	Between		0.153	n=182
	Within		0.028	

Notes: N – total number of observations for each variable; n – number of observed countries for each variable.

independent variables of the models are lagged by two periods. Variance inflated factors are calculated at each stage of the analysis to control for the possibility of multicollinearity and robust standard errors are also used to avoid heteroscedasticity. At last, for purposes of coherency and to facilitate the interpretation of the regression coefficients, all the indexes that are originally in a different scale were rescaled to a 0 to 1 scale. For all the indexes that are used, 0 represents the lowest possible score and 1 the highest possible score. Therefore, in each of those indexes the value of 1 is obtained by the countries with the most democratic institutions (*autocracy-democracy*), the highest trust in politicians (*trust_politicians*), the highest corruption levels (*corruption*), the highest freedom of the press (*press_free*), the highest level of government effectiveness (*gov_effectiveness*), the highest human development (*HDI*) or the highest level of economic globalization (*eco_global*).

Within the empirical framework mentioned above, the procedure can be divided in three main steps: first, cross section regressions for the different years to which the EPI is available in the sample period are estimated; second, panel regressions are considered; third, the interactive hypotheses are tested, using panel data models that consider interaction terms between the political-institutional variables and either the technology penetration or the human development variables.

A general model that summarizes the first step of the approach is represented by equation (1):

$$EPI_{i,t} = \beta_0 + \beta_1.\text{Institutional}_{i,t-2} + \beta_2.\text{mobile\_cellular}_{i,t-2} + \beta_3.\text{HDI}_{i,t-2} + \gamma.\text{Control}'_{i,t-2} + \varepsilon_{it} \tag{1}$$

where $EPI_{i,t}$ represents the EPI index of country i in the year t, $\text{Institutional}_{i,t-2}$ is a two-period lagged variable that proxies one of the proxies for the five political institutional dimensions considered in the theoretical model. $\text{mobile\_cellular}_{i,t-2}$ relates to H2 and stands for the two-period lagged mobile cellular penetration rate, while $\text{HDI}_{i,t-2}$ relates to H3, standing for the two-period lagged value of the human development index. $\text{Control}'_{i,t-2}$ is a vector of two-period lagged control variables that contains the remaining economic and demographic variables. Finally, ε_{it} stands for the error term, while β_0, β_1, β_2, β_3 and γ represent the parameters, or vectors of parameters, to be estimated.

In the second step, six observations in time (2008, 2010, 2012, 2014, 2016 and 2018) are used, and controls for time effects and country-level fixed effects are added to the model, as represented in equation (2).

$$EPI_{i,t} = \beta_0 + \beta_1.\text{Institutional}_{i,t\text{-}2} + \beta_2.\text{mobile\_cellular}_{i,t\text{-}2} + \beta_3.\text{HDI}_{i,t\text{-}2} + \gamma.\text{Control'}_{i,t\text{-}2} + \lambda_t + \mu_i + \varepsilon_{it}$$

$$(2)$$

In this case, λ_t represents time effects, defined as a set of year dummy variables, μ_i stands for country-level fixed effects and everything else remains as in equation (1). To decide whether we include fixed effects or not, we rely on Hausman specification tests (Hausman, 1978). The null hypothesis of the test is the absence of significant differences between the coefficients of a consistent estimator, the fixed effects one, and an alternative efficient estimator, typically the random effects one. The rejection of the null indicates that the inclusion of fixed effects is needed.[2] As, to be best of the authors' knowledge, there is no standard way of including random effects in the fractional probit framework, a model with no fixed nor random effects was used in the cases of no rejection of the Hausman test's null. In such cases, the results obtained via the fractional regression and the random effects estimates were compared to make sure that the absence of random effects did not substantially affected the results.

To test the interactive hypothesis, extensions of the model represented in equation (2) are estimated. These consider either interactions between the institutional and the technology penetration variable (H4), or interactions between the first and human development (H5). To test H4, the following extension was considered:

$$\begin{aligned} EPI_{i,t} = &\ \beta_0 + \beta_1.\text{Institutional}_{i,t\text{-}2} + \beta_2.\text{mobile\_cellular}_{i,t\text{-}2} \\ &+ \beta_3.\text{Institutional*mobile\_cellular}_{i,t\text{-}2} + \beta_4.\text{HDI}_{i,t\text{-}2} + \gamma.\text{Control'}_{i,t\text{-}2} + \lambda_t + \mu_i + \varepsilon_{it} \end{aligned}$$

$$(3)$$

where $\text{Institutional*mobile\_cellular}_{i,t-2}$ represents the interaction between the institutional and the technology penetration-related variable and everything else remains as in equation (2).

Equation (4) represents the extension that allows to test H5:

$$\begin{aligned} EPI_{i,t} = &\ \beta_0 + \beta_1.\text{Institutional}_{i,t\text{-}2} + \beta_2.\text{HDI}_{i,t\text{-}2} + \beta_3.\text{Institutional*HDI}_{i,t\text{-}2} \\ &+ \beta_4.\text{mobile\_cellular}_{i,t\text{-}2} + \gamma.\text{Control'}_{i,t\text{-}2} + \lambda_t + \mu_i + \varepsilon_{it} \end{aligned}$$

$$(4)$$

where $\text{Institutional*HDI}_{i,t-2}$ represents the interaction between the institutional variables and the human development variable and everything else remains as in equation (2).

FINDINGS

Cross Section Results

As explained previously, the analysis starts by contemplating cross sectional models for the different years of the sample period. When scattering the values for the EPI and the autocracy-democracy index, no linear relationship arises[3], so the chosen specification of the model considers two dummy variables: autocracy-democracy_>p75 is equal to 1 whenever, in a given year, the country's score in the autocracy-democracy index is above the percentile 75 of that year scores' distribution, and 0 otherwise; autocracy-democracy_<p25 is equal to 1 whenever, in a given year, the country's score in the autocracy-democracy index is below the percentile 25 of that year scores' distribution, and 0 otherwise. Table 2 presents the results

Table 2. Average marginal effects of the fractional probit regressions – democracy (H1a) – dependent variable: EPI

Variables	(1) 2018	(2) 2016	(3) 2014	(4) 2012	(5) 2010	(6) 2008
autocracy-democracy_>p75	0.091**	0.045	0.035	-0.003	0.030	0.002
	(0.036)	(0.035)	(0.050)	(0.048)	(0.035)	(0.044)
autocracy-democracy_<p25	-0.012	0.020	-0.008	0.011	-0.030	-0.041
	(0.034)	(0.034)	(0.035)	(0.043)	(0.030)	(0.034)
mobile_cellular/100	-0.036	-0.037	0.010	0.014	-0.106**	0.039
	(0.061)	(0.054)	(0.059)	(0.055)	(0.052)	(0.061)
HDI	0.974***	1.189***	1.091***	1.112***	1.031***	0.622***
	(0.141)	(0.167)	(0.177)	(0.206)	(0.184)	(0.210)
pop (billions)	0.573**	0.310***	0.260***	0.120	0.131***	0.223***
	(0.251)	(0.068)	(0.067)	(0.079)	(0.031)	(0.054)
eco_global	0.239*	0.101	0.039	0.010	0.091	0.055
	(0.137)	(0.132)	(0.133)	(0.148)	(0.118)	(0.149)
Observations	158	157	157	158	156	156
Pseudo R2	0.162	0.144	0.126	0.169	0.148	0.108
Log-likelihood	-88.72	-93.12	-93.97	-74.23	-70.24	-72.92
AIC	191.4	200.2	201.9	162.5	154.5	159.8
SIC	212.9	221.6	223.3	183.9	175.8	181.2

Notes: All models were estimated with a constant. Robust standard errors in parentheses. Statistical significance: *** $p<0.01$, ** $p<0.05$, * $p<0.1$.

for the average marginal effects of the fractional probit regressions of the models that follow equation (1) and include the variables related to democracy (H1a), with each column corresponding to a different year.

The results reveal that the more democratic countries are, when compared to countries whose EPI is between the sample percentile 25 and 75, associated with higher EPI scores. However, that is only true for the last year of the sample. In contrast no significant results were obtained for the dummy variable that identifies the most autocratic countries (autocracy-democracy_<p25).

Regarding the results for the remaining variables, strong support for H3 was found. The HDI's coefficient is positive and significant for all the years. Moreover, the magnitude of the coefficient reveals to be stable, except for the year of 2008, where the coefficient drops from a value around 1 to approximately 0.6. Population also exerts a positive and significant impact on EPI, and only for 2012 its coefficient is not significant. In contrast, no support for H2 is found, neither for the relationship between the level of economic globalization and e-participation.

Table 3. Average marginal effects of the fractional probit regressions – trust (H1b) – dependent variable: EPI

	(1)	(2)	(3)	(4)	(5)	(6)
Variables	**2018**	**2016**	**2014**	**2012**	**2010**	**2008**
trust_politicians	0.092	0.034	0.076	0.237**	0.078	0.118
	(0.089)	(0.086)	(0.115)	(0.098)	(0.100)	(0.083)
mobille_cellular/100	-0.045	-0.068	-0.061	0.015	-0.125*	0.034
	(0.058)	(0.054)	(0.060)	(0.061)	(0.068)	(0.066)
HDI	1.009***	1.212***	1.151***	1.127***	1.168***	0.627***
	(0.133)	(0.169)	(0.168)	(0.207)	(0.216)	(0.232)
pop (billions)	0.466**	0.294***	0.244***	0.092	0.131***	0.233***
	(0.195)	(0.064)	(0.078)	(0.092)	(0.042)	(0.058)
eco_global	0.208	0.136	0.020	-0.156	0.064	0.035
	(0.139)	(0.155)	(0.147)	(0.179)	(0.167)	(0.171)
Observations	136	140	141	136	129	148
Pseudo R2	0.141	0.122	0.0920	0.128	0.116	0.0944
Log-likelihood	-74.48	-84.65	-88.48	-71.78	-64.68	-72.11
AIC	161	181.3	189	155.6	141.4	156.2
SIC	178.4	199	206.6	173	158.5	174.2

Notes: All models were estimated with a constant. Robust standard errors in parentheses. Statistical significance: *** $p<0.01$, ** $p<0.05$, * $p<0.1$.

The results for the average marginal effects of the fractional probit regressions of the models that follow equation (1) and include the variables related to public trust in politicians are presented in Table 3. Once again, each column corresponds to a different year.

The cross-sectional results do not provide a strong support to H1b. In spite of always exhibiting positive coefficients, meaning that a higher public trust in politicians is associated with higher e-participation levels, the *trust_politicians* variable is only significant in the year of 2012. For that year, it is estimated that on average, an increase of one point in the Public Trust in Politicians index is associated with an increase of approximately 0.24 points in the EPI.

Regarding *mobile_cellular* and HDI, the variables related with H2 and H3, the scenario is consistent with the one described for Table 2. The same applies to the two control variables, pop and *eco_global*.

Table 4 presents the results for the models where corruption is an independent variable. In this case, to make the interpretation of the results more intuitive and consistent with the sign predicted in H1c, the scale of the corruption index was inverted, in a way that higher values of the index correspond to higher perceived corruption levels.

As hypothesized, the results point to a negative relationship between corruption and e-participation levels. The coefficients of the corruption variable are significant for two of the six years considered: 2008 and 2018. For those years, it is estimated that, on average, a one-point increase in the corruption index is associated with a 0.26 and 0.31 decrease in the EPI.

All the remaining variables in the model follow the pattern of the previous tables: no support for the importance of technology penetration and economic globalization and strong support for the importance of socioeconomic development and the size of population on predicting e-participation levels.

As in the corruption index case, the scale of the Freedom of the Press index was inverted to make the interpretation of the coefficients easier and consistent with the hypotheses presented earlier in this chapter. Therefore, higher values of the index correspond to higher freedom of the press and positive coefficients associated with *pressfree* mean that more freedom of the press is associated with higher e-participation levels. Table 5 presents the results for the average marginal effects of the fractional probit regressions of the models that include the variable related to press freedom (H1d).

The results reveal that, although mostly positive as hypothesized, the coefficients of the *pressfree* variable are never significant. Therefore, H1d is not supported. Regarding the remaining four independent variables of the model, nothing substantially new arises when comparing the results of Table 5 with the ones reported in Tables 2 to 4.

Table 4. Average marginal effects of the fractional probit regressions – corruption (H1c) – dependent variable: EPI

Variables	(1) 2018	(2) 2016	(3) 2014	(4) 2012	(5) 2010	(6) 2008
corruption	-0.305***	-0.137	-0.171	-0.110	-0.087	-0.261***
	(0.111)	(0.100)	(0.112)	(0.103)	(0.084)	(0.101)
mobile_cellular/100	0.034	0.019	0.018	0.014	-0.085*	0.021
	(0.054)	(0.048)	(0.050)	(0.050)	(0.047)	(0.056)
HDI	0.827***	1.057***	0.934***	0.959***	0.898***	0.352*
	(0.155)	(0.169)	(0.173)	(0.203)	(0.173)	(0.187)
pop (billions)	0.685**	0.383***	0.313***	0.156*	0.163***	0.246***
	(0.344)	(0.093)	(0.086)	(0.083)	(0.041)	(0.063)
eco_global	0.097	0.029	0.010	-0.021	0.115	0.075
	(0.137)	(0.130)	(0.136)	(0.143)	(0.120)	(0.140)
Observations	180	179	179	180	180	179
Pseudo R2	0.143	0.125	0.109	0.155	0.135	0.107
Log-likelihood	-104.6	-108.5	-108	-82.80	-77.98	-80.19
AIC	221.1	229	228	177.6	168	172.4
SIC	240.3	248.1	247.1	196.8	187.1	191.5

Notes: All models were estimated with a constant. Robust standard errors in parentheses. Statistical significance: *** $p<0.01$, ** $p<0.05$, * $p<0.1$.

Lastly, Table 6 presents the results for the average marginal effects of the fractional probit regressions of the models that include the variable related to government effectiveness (H1e). As in the remaining tables of this section, each column corresponds to a different year.

The hypothesis that higher government effectiveness is associated with higher e-participation levels (H1e) is strongly supported by the results. The coefficients associated with gov_effectiveness are positive and significant for all the years under studied. It is also worth mentioning that the magnitude of the coefficients exhibits a positive trend over time. From 2008 to 2014, it is estimated that, on average, a one-point increase in the government effectiveness index, is associated with an increase of 0.28 to 0.4 points in the EPI. However, in the most recent years, 2016 and 2018, the estimated coefficients are respectively around 0.59 and 0.7.

Once again, the results for the remaining independent variables are similar to the ones reported in Tables 2 to 5.

Table 5. Average marginal effects of the fractional probit regressions – freedom of the press (H1d) – dependent variable: EPI

	(1)	(2)	(3)	(4)	(5)	(6)
Variables	**2018**	**2016**	**2014**	**2012**	**2010**	**2008**
pressfree	0.255	0.097	0.040	-0.048	0.095	0.175
	(0.161)	(0.165)	(0.076)	(0.110)	(0.094)	(0.112)
mobile_cellular/100	-0.012	-0.019	0.013	0.017	-0.112**	0.013
	(0.057)	(0.052)	(0.054)	(0.053)	(0.052)	(0.064)
HDI	0.980***	1.186***	1.097***	1.096***	1.024***	0.642***
	(0.132)	(0.150)	(0.153)	(0.174)	(0.173)	(0.206)
pop (billions)	0.662**	0.348***	0.285***	0.126	0.153***	0.250***
	(0.302)	(0.076)	(0.075)	(0.084)	(0.038)	(0.066)
eco_global	0.237*	0.104	0.021	0.011	0.123	0.021
	(0.137)	(0.127)	(0.132)	(0.147)	(0.123)	(0.160)
Observations	166	164	164	165	161	156
Pseudo R2	0.149	0.136	0.114	0.159	0.138	0.102
Log-likelihood	-94.82	-98.25	-99.21	-77.85	-72.40	-72.90
AIC	201.6	208.5	210.4	167.7	156.8	157.8
SIC	220.3	227.1	229	186.3	175.3	176.1

Notes: All models were estimated with a constant. Robust standard errors in parentheses. Statistical significance: *** $p<0.01$, ** $p<0.05$, * $p<0.1$.

Panel Results

The second step of the empirical analysis considers panel regressions, with six observations in time (2008, 2010, 2012, 2014, 2016 and 2018). Time effects and, in some cases, fixed effects, were added, as described by equation (2) above. As high variance inflated factors were found for the *eco_global* variable and it was almost never statistically significant in the cross-sectional regressions, it is excluded here. Table 7 contains the results for the average marginal effects of the fractional probit regressions. Column (1) includes the democracy-related dummy variables, column (2) the public trust in politicians index, column (3) the corruption index, column (4) the freedom of the press index and column (5) the government effectiveness one.

Regarding the hypothesis related with the political institutional variables, the panel results are consistent with H1a, H1c and H1e, but not with H1b and H1d. From column (1), it is possible to observe that it is estimated that countries on the top of the distribution of the autocracy-democracy index are, on average, associated with an increase of approximately 0.04 points in the EPI. In turn, a one-point

Table 6. Average marginal effects of the fractional probit regressions – government effectiveness (IIIe) – dependent variable: EPI

Variables	(1) 2018	(2) 2016	(3) 2014	(4) 2012	(5) 2010	(6) 2008
gov_effectiveness	0.696***	0.587***	0.360***	0.325**	0.279***	0.396***
	(0.142)	(0.130)	(0.136)	(0.133)	(0.095)	(0.124)
mobile_cellular/100	0.035	0.032	0.016	0.018	-0.093**	0.008
	(0.049)	(0.043)	(0.047)	(0.049)	(0.046)	(0.056)
HDI	0.473***	0.660***	0.787***	0.759***	0.753***	0.260
	(0.166)	(0.178)	(0.174)	(0.199)	(0.157)	(0.177)
pop (billions)	0.545**	0.316***	0.281***	0.122	0.130***	0.209***
	(0.270)	(0.077)	(0.081)	(0.078)	(0.040)	(0.059)
eco_global	-0.021	-0.136	-0.076	-0.113	0.036	0.022
	(0.133)	(0.126)	(0.141)	(0.153)	(0.122)	(0.145)
Observations	180	179	179	180	180	179
Pseudo R2	0.153	0.134	0.112	0.161	0.141	0.110
Log-likelihood	-103.3	-107.4	-107.6	-82.25	-77.48	-79.86
AIC	218.6	226.7	227.3	176.5	167	171.7
SIC	237.8	245.9	246.4	195.6	186.1	190.9

Notes: All models were estimated with a constant. Robust standard errors in parentheses. Statistical significance: *** $p<0.01$, ** $p<0.05$, * $p<0.1$.

Table 7. Average marginal effects of the panel fractional probit regressions – dependent variable: EPI

Variables	(1) democracy	(2) trust	(3) corruption	(4) press freedom	(5) gov_effec
autocracy-democracy_<p25	-0.016				
	(0.014)				
autocracy-democracy_>p75	0.038**				
	(0.017)				
trust_politicians		0.006			
		(0.065)			
corruption			-0.176***		
			(0.040)		
pressfree				-0.069	
				(0.055)	

continues on following page

Table 7. Continued

	(1)	(2)	(3)	(4)	(5)
gov_effectiveness					0.380***
					(0.048)
mobile_cellular/100	0.010	-0.014	0.037*	-0.025	0.029
	(0.022)	(0.029)	(0.019)	(0.027)	(0.019)
HDI	1.002***	0.766	0.788***	0.363	0.561***
	(0.062)	(0.551)	(0.065)	(0.426)	(0.067)
pop (billions)	0.232***	0.701	0.285***	0.839	0.256***
	(0.032)	(0.558)	(0.038)	(0.522)	(0.037)
year: 2010	-0.013	-0.006	-0.018	0.001	-0.014
	(0.019)	(0.017)	(0.018)	(0.014)	(0.018)
year: 2012	0.009	0.032	0.004	0.030*	0.010
	(0.022)	(0.020)	(0.021)	(0.017)	(0.020)
year: 2014	0.179***	0.213***	0.168***	0.206***	0.176***
	(0.023)	(0.024)	(0.021)	(0.021)	(0.020)
year: 2016	0.240***	0.283***	0.228***	0.283***	0.240***
	(0.023)	(0.028)	(0.021)	(0.023)	(0.020)
year: 2018	0.332***	0.384***	0.325***	0.381***	0.339***
	(0.024)	(0.031)	(0.022)	(0.026)	(0.021)
Observations	948	830	1,095	986	1,095
# of countries	160	148	184	170	184
Hausman statistic	9.29	19.55	9.60	29.81	7.83
Fixed Effects	No	Yes	No	Yes	No
Pseudo R2	0.202	0.253	0.189	0.268	0.194
Log-likelihood	-499.2	-419.5	-575.6	-476.3	-572
AIC	1020	1153	1171	1311	1164
SIC	1074	1894	1221	2187	1214

Notes: All models were estimated with a constant. Robust standard errors in parentheses. 2008 is the base category of the set of year dummy variables. Statistical significance: *** $p<0.01$, ** $p<0.05$, * $p<0.1$.

decrease in the corruption index is estimated to be associated with an increase of approximately 0.18 points in the EPI. Lastly, one additional point in the government effectiveness index is estimated to be associated with an increase of 0.38 points in the e-participation score.

Unlike in the cross-sectional regressions, where no support was found for H2, column (3) of Table 7 reports some anecdotal evidence of a possible impact of the technology penetration in the EPI. Regarding socioeconomic development (H3), positive and significant results are found in columns (1), (3) and (5), but not in columns (2) and (4), the model where fixed effects are used. A similar landscape is found for population, with no significant results in the models that use fixed effects, but positive and significant results in the remaining models. In both cases, it is not the magnitude of the coefficient that drops dramatically when fixed effects are included; it is the standard error that increases. Recalling the descriptive statistics of Table 1, it is likely that such occurrence is explained by the low within variation that both HDI and *pop* exhibit along the sample period. Finally, the results for the year dummy variables point to a global increase in the EPI levels in the most recent years of the sample. From 2014 onwards, all the coefficients associated with these variables display positive and significant coefficients, following the pattern of a positive trend.

Interaction Terms

Interactions Between Institutions and Technology Penetration

In this subsection, we report the results of the models that were estimated to test H4, the interactive hypothesis that posits that the impact that political institutional factors exert on the e-participation levels may vary according to the sophistication of the technology penetration.

Table 8 presents the fractional probit regression coefficients for six different models that follow the previously presented equation (3). Columns (1) and (2) are related to the interactions between the autocracy-democracy dummies and the mobile cellular penetration. The remaining columns present, in this order, the results for the interactions between *mobile_cellular* and public trust in politicians, corruption levels, freedom of the press and government effectiveness. For reasons of parsimony, only the results for the variables involved in the interaction terms are presented.[4]

The results presented in Table 8 reveal that the interactions between the democracy and the technology penetration-related variables are statistically significant. The same happens for the interactions between the latter and the corruption index, as well as the government effectiveness index. Therefore, the evidence suggests that the impact that democratic institutions, corruption levels and government effectiveness exert in the e-participation levels varies according to the technology penetration in each country.

To get additional information about how technology penetration mediates the relationship between the institutional variables and e-participation, the average marginal effects of the institutional variables on the EPI along the *mobile_cellular* distribution were plotted. Figure 3 presents the plots for the four interactions terms

Table 8. Fractional probit regressions coefficients including interaction terms –
dependent variable: EPI

Variables	(1) democracy	(2) democracy	(3) trust	(4) corruption	(5) press freedom	(6) gov_effec
autocracy-democracy_<p25	-0.220**	-0.052				
	(0.099)	(0.049)				
autocracy-democracy_<p25* wdi_mobile/100	0.203**					
	(0.101)					
autocracy-democracy_>p75	0.141**	0.230				
	(0.058)	(0.158)				
autocracy-democracy_>p75* wdi_mobile/100		-0.093				
		(0.129)				
trust_politicians			-0.103			
			(0.312)			
trust_politicians* wdi_mobile/100			0.158			
			(0.276)			
corruption				0.031		
				(0.308)		
corruption* wdi_mobile/100				-0.604**		
				(0.271)		
pressfree					0.043	
					(0.332)	
pressfree* wdi_mobile/100					-0.357	
					(0.333)	
gov_effectiveness						0.797**
						(0.315)
gov_effectiveness* wdi_mobile/100						0.471*
						(0.268)
wdi_mobile/100	-0.046	0.045	-0.119	0.445***	0.159	-0.119
	(0.080)	(0.078)	(0.155)	(0.159)	(0.273)	(0.135)
Observations	948	948	830	1,095	986	1,095
# of countries	160	160	148	184	170	184
Fixed effects	No	No	Yes	No	Yes	No
Pseudo R2	0.203	0.202	0.253	0.190	0.268	0.195
Log-likelihood	-498.9	-499.2	-419.5	-575.2	-476.3	-571.8

continues on following page

Table 8. Continued

Variables	(1) democracy	(2) democracy	(3) trust	(4) corruption	(5) press freedom	(6) gov_effec
AIC	1022	1022	1155	1172	1313	1166
SIC	1080	1081	1901	1227	2193	1221

Notes: All models were estimated with a constant. Robust standard errors in parentheses. Year dummies included: 2008 is the base category of the set of year dummy variables. HDI and *pop* as additional independent variables. Statistical significance: *** $p<0.01$, ** $p<0.05$, * $p<0.1$.

where some regions of statistical significance were found.[5] The black line inside the blue area represents the estimated average marginal, effects. The blue area represents the 95% confidence interval. The red vertical line stands for the mean value of *mobile_cellular* over the entire sample period.

The upper left plot presents the average marginal effects of the variable autocracy-democracy_<p25 along the *mobile_cellular* distribution. It reveals that harsh autocracies result in poorer EPI scores, but only when technology penetration is low. On the contrary, the upper right plot reveals that the positive effect that solid democracies may have on the EPI is only valid for values around the mean of *mobile_celullar*. Both plots point to the idea that the democratic degree of the institutions is neutral in contexts of higher technology penetration. In the lower left plot, it is possible to observe that the negative effect on EPI associated with high levels of corruption is stronger when technology penetration is higher. At last, the lower right plot indicates that, although always positive and significant, the average marginal effect of *gov_effectiveness* on e-participation is higher when technology penetration is higher.

Interactions Between Institutions and Socioeconomic Development

H5 postulates that the impact that political institutional factors exert on e-participation may vary according to the socio-economic development of each country. Following equation (4), Table 9 presents the results for six models that consider interaction terms between the institutional variables and the human development index. The first two columns report the interactions between the autocracy-democracy dummies and HDI. Columns (3) to (6) present, in this order, the results for the interactions between the human development index and public trust in politicians, corruption levels, freedom of the press and government effectiveness. As in the table of the previous subsection, for reasons of parsimony, only the results for the variables involved in the interaction terms are presented.

Figure 3. Average Marginal Effects of the institutional variables with 95% confidence intervals. Effects on the conditional mean of EPI in the vertical axis. Values of mobile_cellular in the horizontal axis. Mean value of mobile_cellular in the red vertical line.

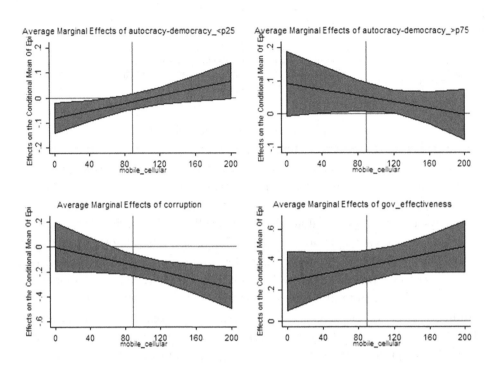

Table 9. Fractional probit regressions coefficients including interaction terms – dependent variable: EPI

Variables	(1) democracy	(2) democracy	(3) trust	(4) corruption	(5) press freedom	(6) gov_effec
autocracy-democracy_<p25	0.288	-0.049				
	(0.237)	(0.048)				
autocracy-democracy_<p25* HDI	-0.507					
	(0.351)					
autocracy-democracy_>p75	0.104*	-1.063**				
	(0.061)	(0.492)				
autocracy-democracy_>p75* HDI		1.417**				
		(0.586)				

continues on following page

Table 9. Continued

Variables	(1) democracy	(2) democracy	(3) trust	(4) corruption	(5) press freedom	(6) gov_effec
trust_politicians			-0.999			
			(0.912)			
trust_politicians* HDI			1.465			
			(1.293)			
corruption				1.945***		
				(0.614)		
corruption* HDI				-3.122***		
				(0.726)		
pressfree					2.893***	
					(0.926)	
pressfree* HDI					-4.582***	
					(1.336)	
gov_effectiveness						-0.349
						(0.608)
gov_effectiveness* HDI						1.961***
						(0.703)
HDI	3.463***	3.211***	2.198	4.386***	4.536***	1.104***
	(0.234)	(0.227)	(1.989)	(0.497)	(1.716)	(0.339)
Observations	948	948	830	1,095	986	1,095
# of countries	160	160	148	184	170	184
Fixed effects	No	No	Yes	No	Yes	No
Pseudo R2	0.203	0.203	0.253	0.192	0.268	0.195
Log-likelihood	-499	-498.7	-419.4	-574	-475.9	-571.4
AIC	1022	1021	1155	1170	1312	1165
SIC	1080	1080	1901	1225	2193	1220

Notes: All models were estimated with a constant. Robust standard errors in parentheses. Year dummies included: 2008 is the base category of the set of year dummy variables. *Mobile_cellular* and *pop* as additional independent variables. Statistical significance: *** $p<0.01$, ** $p<0.05$, * $p<0.1$.

From Table 9, it is possible to observe that the results support H5. In particular, the interactions between the HDI and the high democracy score dummy, as well as with the corruption index, the freedom of the press, and government effectiveness are statistically significant.

As in the previous section, the following Figure presents the plots for the terms where regions of statistically significant average marginal effects of the institutional variable on the EPI along the distribution of the HDI were found. Once again, the black line inside the blue area represents the estimated average marginal effects, while the blue area represents the 95% confidence interval. The red vertical line stands for the mean value of HDI over the entire sample period.

Figure 4. Average Marginal Effects of the institutional variables with 95% confidence intervals. Effects on the conditional mean of EPI in the vertical axis. Values of HDI in the horizontal axis. Mean value of HDI in the red vertical line.

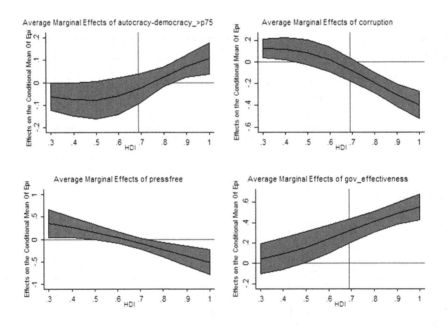

The upper left plot presents the average marginal effects of the variable autocracy-democracy_>p75. It reveals that the positive effect that is found for solid democracies on the EPI only holds for contexts in which the socioeconomic development is high. The upper right plot stands for the average marginal effects of corruption. Regarding the corrosive effect that corruption may have on the EPI only occurs when socioeconomic development is above the mean. In fact, there is even anecdotal evidence of the contrary for very low values of the HDI. Despite no significant relationship being found between freedom of the press and e-participation in the previous sections, the lower left graph provides anecdotal evidence that freedom of the press may exert a positive effect in contexts where socioeconomic development is low, and work the other way around when socioeconomic development is high. Finally, as in the technology penetration case, the lower right plot suggests that governmental effectiveness exerts a higher positive impact on e-participation when socioeconomic development is higher.

DISCUSSION

The overall picture emerging from the findings confirms and extends prior studies on the determinants of e-participation. Among the variables assessing the quality of political institutions, government effectiveness is the strongest predictor of higher EPI levels, thus confirming the idea expressed in Gulati *et al.* (2014) that higher professionalization of public administration supports the adoption and implementation of e-participation tools. The other variables addressing different aspects of the quality of political institutions are less consistent over time and only appear as relevant predictors in the panel model. Nevertheless, they confirm the hypothesized relationships: higher EPI levels appear in countries characterized by better democratic performance, freedom of the press, and lower corruption levels. Overall, the findings indicate that the quality of political institutions is a crucial contextual element to nurture e-participation initiatives.

Another important finding of the analyses included in this chapter is the rejection of technological determinism when it comes to e-participation (Susha & Grönlund, 2012). Better technological penetration, as measured by mobile cellular phone subscriptions per 100 citizens in a country, does not appear to be associated with higher e-participation levels. If anything, there is a quality threshold beyond which technology penetration is unrelated to e-participation. More importantly, the results show that technology penetration mediates the relationship between several indicators of the quality of political institutions. First, higher mobile penetration reinforces the positive association between government effectiveness and e-participation levels, which is consistent with the idea that technological access is important in taking advantage of effective public sectors promoting electronic participation tools. Second, the finding that more corrupt countries also display lower levels of e-participation is not surprising in itself. However, the idea that better technological penetration has a dampening effect on this relationship is discouraging, since it suggests that technology may actually contribute to deepen the already negative effects of corruption. Finally, as with prior empirical work, the evidence presented here regarding the interactions of democratic performance, technology penetration and EPI levels is unclear. The worst autocracies with low mobile penetration display the lowest EPI levels, but beyond that the evidence becomes mixed. The empirical analysis does not provide incontrovertible support to the argument advanced by Stier (2015) for e-government performance that autocracies with better technology penetration perform better in the EPI, but it does suggest that this scenario is more likely than the opposite one. In other words, technology penetration levels are likely relevant for the relationship between a country's placement in the autocracy-democracy continuum and its EPI level.

These findings contrast with the result for the socio-economic development variable. The HDI is an important predictor in every single-year specification and in all but two of the panel models. More importantly, the interactive terms support the theoretical argument that e-participation is most successful in countries which have high quality political institutions and higher socio-economic development simultaneously. While this is not exactly a surprising result, the fact that the effect is true for four out of five measures of quality of political institutions is quite remarkable. Socio-economic development also reinforces the expected positive effects of higher democratic performance, lower corruption levels and better government effectiveness on EPI levels.

CONCLUSION

This chapter employed data from the E-Participation Index (EPI) developed by the United Nations Department of Economic and Social Affairs (UNDESA) to analyze the role of the quality of political institutions in promoting e-participation over the period of 2008-2018. The findings indicate that countries with better democratic performance, lower corruption levels and higher government effectiveness are associated with higher EPI scores. While these results are not entirely robust to all model specifications and all years under analysis, they are largely supportive of the argument that better political institutions contribute to promote more electronic participation at the country level.

The results also support the main argument included in the theoretical model that this positive effect of the quality of political institutions is mediated by more contextual factors, such as technology penetration and socio-economic development. Concretely, socio-economic development reinforces this positive effect of the quality of institutions, which reaches the strongest impact in countries with higher HDI scores. The mediating effect of technology penetration, while present, it is far less evident and more mixed. Technology penetration enhances the positive impact of government effectiveness and the absence of corruption on e-participation, but no clearly discernible trend is present in its interaction with the remaining indicators of the quality of political institutions.

Given the set of findings reported in this chapter, national governments aiming to promote the 2030 UN Agenda for Sustainable Development Goals (SDG's) will have to consider additional efforts into the adoption of e-participation tools capable of enhancing the availability of information, involving citizens in broad consultation processes and promoting deliberative decision-making. The successful implementation of these initiatives and the outcomes they are likely to generate will be crucial, not only for the legitimacy goals of elected officials but also to accomplish the ambitious sustainable development goals. Congruent and concerted national agendas for e-participation, while a core concept for e-democracy, shall be considered. The ultimate objective will thus be to contribute for the achievement of goals proclaimed in SDG 16, namely helping in the reduction of corruption, enhancement of transparency and accountability of institutions, promotion of inclusive and participatory processes and policies, and strengthening of good governance principles.

Limitations

The analyses included in this chapter may suffer from a number of limitations, primarily related to the nature of the dependent variable: the E-Participation Index. First, the EPI is questioned on the grounds of validity issues discussed at length in Lidén (2015). However, Lidén's piece assumes the EPI is a measure of e-Democracy and that e-Democracy and e-Participation can be conflated. Given the content of the EPI, this is not an accurate assumption. The second problem relates to the concept of e-participation itself. The EPI does not include outcomes, so the scores may be the result of a search for legitimacy on the part of elected officials rather than a genuine goal of improving e-participation, particularly in authoritarian regimes. Lastly, the analysis is focused on the EPI as a whole, not considering its different dimensions, namely the three main components of the index. This may also be a direction for future research despite of it being contingent on and constrained by the availability of more detailed data.

Another set of limitations relates to the independent variables, particularly those aimed at measuring the quality of political institutions. There is a high persistency on the values of the institutional variables within countries. Institutions typically change slowly and a sample period of ten years, while longer than what most (or all) the previous studies have considered, it is still limited to measure institutional change. A higher variability and a longer sample period would benefit the robustness of the statistical inference and make it more accurate in providing a causal interpretation of the results. Additionally, the range of variables to be considered when conceptualizing the quality of political institutions might be seen as a limitation. They are representative measures to assess the quality of political institutions but are neither exhaustive nor exclusive.

Directions for Future Research

The richness of the panel data included in this chapter should allow the expansion of this comparative analysis to consider different dimensions of e-participation and/ or the regional variation of the EPI country scores. Pending data availability, future research can also investigate the adoption (or the "demand side") of e-participation tools.

This study identifies broad trends in e-participation across the globe based on single country scores. However, as discussed above, the EPI is not without its limitations, so these tendencies need to be explored with more in-depth analyses through regional comparisons and country case studies. Without these more fine-grained efforts, it is likely that the picture of the country trends in e-participation will be incomplete at best.

ACKNOWLEDGMENT

This paper is a result of the project "SmartEGOV: Harnessing EGOV for Smart Governance (Foundations, methods, Tools) / NORTE-01-0145-FEDER-000037", supported by Norte Portugal Regional Operational Programme (NORTE 2020), under the PORTUGAL 2020 Partnership Agreement, through the European Regional Development Fund (EFDR). António Tavares acknowledges the financial support of the Portuguese Foundation for Science and Technology and the Portuguese Ministry of Education and Science through national funds [Grant No. UID/CPO/0758/2019].

REFERENCES

Anduiza, E., Gallego, A., & Cantijoch, M. (2010). Online political participation in Spain: The impact of traditional and internet resources. *Journal of Information Technology & Politics*, 7(4), 356–368. doi:10.1080/19331681003791891

Åström, J., Karlsson, M., Linde, J., & Pirannejad, A. (2012). Understanding the Rise of e-Participation in Non-democracies: Domestic and International Factors. *Government Information Quarterly*, 29(2), 142–150. doi:10.1016/j.giq.2011.09.008

Becker, G. S. (1964). *Human capital: a theoretical and empirical analysis, with special reference to education.* University of Illinois at Urbana-Champaign: Academy for Entrepreneurial Leadership Historical Research Reference in Entrepreneurship.

Bertot, J. C., Jaeger, P. T., & Grimes, J. M. (2012). Promoting Transparency and Accountability through ICTs, social media, and collaborative e-government. *Transforming Government: People. Process and Policy*, *6*, 78–91.

Bussell, J. (2011). Explaining Cross-National Variation in Government Adoption of New Technologies. *International Studies Quarterly*, *55*(1), 267–280. doi:10.1111/j.1468-2478.2010.00644.x

DiMaggio, P., Hargittai, E., Neuman, W. R., & Robinson, J. P. (2001). Social Implications of the Internet. *Annual Review of Sociology*, *27*(1), 307–336. doi:10.1146/annurev.soc.27.1.307

Dreher, A., Gaston, N., & Martens, P. (2008). *Measuring Globalisation: Gauging Its Consequences.* New York: Springer. doi:10.1007/978-0-387-74069-0

Egorov, G., Guriev, S., & Sonin, K. (2009). Why Resource-poor Dictators Allow Freer Media: A Theory and Evidence from Panel Data. *The American Political Science Review*, *103*(4), 645–668. doi:10.1017/S0003055409990219

Gibson, R., Römmele, A., & Williamson, A. (2014). Chasing the Digital Wave: International Perspectives on the Growth of Online Campaigning. *Journal of Information Technology & Politics*, *11*(2), 123–129. doi:10.1080/19331681.2014.903064

Gulati, G. J., Williams, C. B., & Yates, D. J. (2014). Predictors of On-line Services and e-Participation: A Cross-national Comparison. *Government Information Quarterly*, *31*(4), 526–533. doi:10.1016/j.giq.2014.07.005

Hausman, J. A. (1978). Specification tests in econometrics. *Econometrica*, *46*(6), 1251–1271. doi:10.2307/1913827

Hollyer, J. R., Rosendorff, B. P., & Vreeland, J. R. (2014). Measuring Transparency. *Political Analysis*, *22*(4), 413–434. doi:10.1093/pan/mpu001

Ifinedo, P. (2012). Drivers of E-Government Maturity in Two Developing Regions: Focus on Latin America and Sub-Saharan Africa. *Journal of Information Systems and Technology Management*, *9*(1), 5–22. doi:10.4301/S1807-17752012000100001

Ifinedo, P., & Singh, M. (2011). Determinants of eGovernment Maturity in the Transition Economies of Central and Eastern Europe. *Electronic Journal of E-Government*, *9*(2), 166–182.

Jho, W., & Song, K. J. (2015). Institutional and Technological Determinants of Civil e-Participation: Solo or Duet? *Government Information Quarterly, 32*(4), 488–495. doi:10.1016/j.giq.2015.09.003

Khoirunnida, H., A. N., Purwandari, B., Kartika, D., & Kosandi, M. (2017). Factors Influencing Citizen's Intention to Participate Electronically: The Perspectives of Social Cognitive Theory and e-Government Service Quality. *2017 International Conference on Advanced Computer Science and Information Systems, ICACSIS 2017*, 166–171. 10.1109/ICACSIS.2017.8355028

King, G., Pan, J., & Roberts, M. (2013). How Censorship in China Allows Government Criticism but Silences Collective Expression. *The American Political Science Review, 107*(2), 326–343. doi:10.1017/S0003055413000014

Kneuer, M., & Harnisch, S. (2016). Diffusion of e-government and e-participation in Democracies and Autocracies. *Global Policy, 7*(4), 548–556. doi:10.1111/1758-5899.12372

Krishnan, S., Teo, T. S. H., & Lymm, J. (2017). Determinants of Electronic Participation and Electronic Government Maturity: Insights from Cross-country Data. *International Journal of Information Management, 37*(4), 297–312. doi:10.1016/j.ijinfomgt.2017.03.002

Larosiliere, G. D., & Carter, L. D. (2016). Using a Fit-viability Approach to Explore the Determinants of e-Government Maturity. *Journal of Computer Information Systems, 56*(4), 271–279. doi:10.1080/08874417.2016.1163995

Lidén, G. (2015). Technology and Democracy: Validity in Measurements of e-Democracy. *Democratization, 22*(4), 698–713. doi:10.1080/13510347.2013.873407

Medaglia, R. (2012). E-Participation Research: Moving Characterization Forward (2006-2011). *Government Information Quarterly, 29*(3), 346–360. doi:10.1016/j.giq.2012.02.010

Milner, H. V. (2006). The Digital Divide: The Role of Political Institutions in Technology Diffusion. *Comparative Political Studies, 39*(2), 176–199. doi:10.1177/0010414005282983

Noesselt, N. (2014). Microblogs and the Adaptation of the Chinese Party-State's Governance Strategy. *Governance: An International Journal of Policy, Administration and Institutions, 27*(3), 449–468. doi:10.1111/gove.12045

Norris, D. F., & Moon, M. J. (2005). Advancing E-Government at the Grassroots: Tortoise or Hare? *Public Administration Review*, *65*(1), 64–75. doi:10.1111/j.1540-6210.2005.00431.x

Noveck, B. S. (2009). *Wiki government: How technology can make government better, democracy stronger, and citizens more powerful*. Washington, DC: Brookings Institution Press.

Novo Vázquez, A., & Rosalía Vicente, M. (2019). Exploring the Determinants of e-Participation in Smart Cities. In M. P. Rodríguez Bolívar & L. Alcaide Muñoz (Eds.), *E-Participation in Smart Cities: Technologies and Models of Governance for Citizen Engagement* (pp. 157–178). doi:10.1007/978-3-319-89474-4_8

OECD. (2003). *Promise and Problems of E-Democracy: Challenges of Online Citizen Engagement*. Retrieved 26 July 2017, from http://www.oecd.org/gov/digital-government/35176328.pdf

Papke, L. E., & Wooldridge, J. M. (1996). Econometric methods for fractional response variables with an application to 401 (k) plan participation rates. *Journal of Applied Econometrics*, *11*(6), 619–632. doi:10.1002/(SICI)1099-1255(199611)11:6<619::AID-JAE418>3.0.CO;2-1

Porumbescu, G. A. (2016). Comparing the effects of e-government and social media use on trust in government: Evidence from Seoul, South Korea. *Public Management Review*, *18*(9), 1308–1334. doi:10.1080/14719037.2015.1100751

Rodríguez Bolívar, M. P. (2018). Creative Citizenship: The New Wave for Collaborative Environments in Smart Cities. *Academia (Caracas)*, *31*(1), 277–302. doi:10.1108/ARLA-04-2017-0133

Saebø, Ø., Rose, J., & Flak, L. S. (2008). The shape of eParticipation: Characterizing an emerging research area. *Government Information Quarterly*, *25*(3), 400–428. doi:10.1016/j.giq.2007.04.007

Schultz, T. W. (1961). Investment in human capital. *The American Economic Review*, *51*(1), 1–17.

Scott, J. K. (2006). "E" the People: Do U.S. Municipal Government Web Sites Support Public Involvement? *Public Administration Review*, *66*(3), 341–353. doi:10.1111/j.1540-6210.2006.00593.x

Siau, K., & Long, Y. (2009). Factors Impacting E-Government Development. *Journal of Computer Information Systems*, *50*(1), 98–107. doi:10.1080/08874417.2009.11645367

Singh, H., Das, A., & Joseph, D. (2007). Country-Level Determinants of E-Government Maturity. *Communications of the Association for Information Systems, 20*, 632–648. doi:10.17705/1CAIS.02040

Stier, S. (2015). Political Determinants of e-Government Performance Revisited: Comparing Democracies and Autocracies. *Government Information Quarterly, 32*(3), 270–278. doi:10.1016/j.giq.2015.05.004

Susha, I., & Grönlund, Å. (2012). eParticipation Research: Systematizing the Field. *Government Information Quarterly, 29*(3), 373–382. doi:10.1016/j.giq.2011.11.005

Sylvester, D. E., & McGlynn, A. J. (2010). The Digital Divide, Political Participation, and Place. *Social Science Computer Review, 28*(1), 64–74. doi:10.1177/0894439309335148

United Nations. (2018). *E-Government Survey 2018: Gearing E-Government to support transformation towards sustainable and resilient societies.* Author.

Verba, S., Schlozman, K. L., & Brady, H. (1995). *Voice and equality: Civic voluntarism in American politics.* Cambridge: Cambridge University Press.

Vicente, M. R., & Novo, A. (2014). An Empirical Analysis of e-Participation. The Role of Social Networks and e-Government over Citizens' Online Engagement. *Government Information Quarterly, 31*(3), 379–387. doi:10.1016/j.giq.2013.12.006

Wirtz, B. W., Daiser, P., & Binkowska, B. (2018). E-participation: A Strategic Framework. *International Journal of Public Administration, 41*(1), 1–12. doi:10.1 080/01900692.2016.1242620

Zhang, H., Xu, X., & Xiao, J. (2014). Diffusion of e-Government: A Literature Review and Directions for Future Directions. *Government Information Quarterly, 31*(4), 631–636. doi:10.1016/j.giq.2013.10.013

Zheng, Y., & Schachter, H. L. (2017). Explaining Citizens' E-Participation Use: The Role of Perceived Advantages. *Public Organization Review, 17*(3), 409–428. doi:10.100711115-016-0346-2

Zolotov, M. N., Oliveira, T., & Casteleyn, S. (2018). E-Participation Adoption Models Research in the last 17 Years: A Weight and Meta-analytical Review. *Computers in Human Behavior, 81*, 350–365. doi:10.1016/j.chb.2017.12.031

ENDNOTES

[1]   The survey questionnaire in which the 2018 index is based, was implemented in 2017. The same happens in the remaining years to which the EPI is available.

[2]   To implement the test, auxiliary fixed effects and random effects regressions are estimated. A correction to base both (co)variance matrices on disturbance variance estimate from the efficient estimator is applied whenever the covariance matrix of the test did not reveal to be positive definite and the rank of the differenced variance matrix was equal to the number of coefficients being tested.

[3]   The scatter plot may be provided by the authors upon request. Moreover, when using a panel model with random effects, one of the dummies reveals to be statistically significant, but the same does not happen with the original variable.

[4]   The results for the remaining variables of the six models will be provided by the authors upon request.

[5]   For reasons of parsimony, the remaining two plots are not presented. They will be provided by the authors upon request.

Chapter 6
Civic Technology and Data for Good:
Evolutionary Developments or Disruptive Change in E-Participation?

John G. McNutt
University of Delaware, USA

Lauri Goldkind
Fordham University, USA

ABSTRACT

Governments have long dealt with the issue of engaging their constituents in the process of governance, and e-participation efforts have been a part of this effort. Almost all of these efforts have been controlled by government. Civic technology and data4good, fueled by the movement toward open government and open civic data, represent a sea change in this relationship. A similar movement is data for good, which uses volunteer data scientists to address social problems using advanced analytics and large datasets. Working through a variety of organizations, they apply the power of data to problems. This chapter will explore these possibilities and outline a set of scenarios that might be possible. The chapter has four parts. The first part looks at citizen participation in broad brush, with special attention to e-participation. The next two sections look at civic technology and data4good. The final section looks at the possible changes that these two embryonic movements can have on the structure of participation in government and to the nature of public management.

DOI: 10.4018/978-1-7998-1526-6.ch006

INTRODUCTION

Governments have long dealt with the issue of engaging their constituents in the process of governance. While there have been many attempts to strike the perfect balance between citizen input and management requirements, no perfect system has been developed. Citizen participation and engagement are major concerns of political scientists and public administrators (Verba, Schlozman & Brady, 1995; Schlotzmann, Brady & Verba, 2018) and substantial work has been invested in addressing this need over many years.

E-Participation efforts have been a part of this endeavor. These attempts have met with success in some quarters and have had a less positive impact in others. The emerging developments in smart cities devote considerable effort in how the voice of citizens can be heard in the electronic agora (Desouza & Bhagwatwar, 2012; 2014).

Sadly, not all is well in the virtual town hall. Almost all major e-participation efforts have been at least somewhat controlled by government and at least part of the discussion focuses on ways that government can limit or structure participation rather than promote it. While this makes excellent sense from the perspective of minimizing the effort needed to deal with citizen pressures, but does it really solve the problem? Citizens who do not feel that their voices are heard will not support the government and may very well resort to other means to secure their ends.

New efforts to use the power of technology to promote citizen involvement emerge on a regular basis. While many take the traditional route of soliciting opinions, others move toward a deeper level of involvement.

This chapter discusses two emerging movements that could change the focus of the debate about who can and should control e-participation. They represent a middle ground between e-participation efforts to secure and control public participation and outright alternatives to public efforts. These two movements are *Civic Technology* and *Data for Good*. Both movements are powered by data, technology and the spirit of shared collective intelligence. As such, both have a significant relationship to the rise of open government and governmental transparency (Lathrop & Ruma, 2010) and movements such as Smart Cities. This pushes beyond how many have seen public involvement and brings with it the promise of innovation.

This theoretical chapter will explore these possibilities and their implications for public administration paradigms . The chapter has four parts. The first part looks at citizen participation in broad brush, with special attention to e-participation. The next two sections look at civic technology and data4good. The final section looks at the possible changes that these two embryonic movements can have on the structure of participation in government and to the nature of public management.

Public Management, Citizen Participation and the Growth of E-Participation

The involvement of citizen's Citizen participation and citizen has always been an interesting issue for public management. It reflects the political nature of public management and the relationship between organizations and their environments. Brainard and McNutt (2010) observe that public managers' orientation toward public engagement is informed by the theories that advise their management practice. Their findings support the idea that technology use reflects different theoretical traditions in public administration as expressed by different organizational units. They point to three approaches--Old Public Administration, New Public Management and The New Public Service—that influence public manager's thinking about public participation today.

Traditional Older public administration is based on Weberian concepts of rational management and command and control leadership. It includes ideas from a number of early management theorists (such as Fredrick Taylor, 1937). Political institutions provide the means for public involvement and the bureaucracy carries out the will of the people as expressed by political leaders. This means that involving citizens in the decision-making process is primarily the role of political leaders, not public administrators.

The New Public Management, developed in the 1980s, incorporated ideas from business management and technology (Gruening, 2001; Hood, 1991; Reiter & Klenk, 2019.). This approach sees the public manager as an independent actor that shapes the policy making environment and delivers services based on research. It includes support for performance management, outsourcing and E-government and technology. The degree of public involvement is minimal here as well, although there is some facility for citizen preferences to be considered.

The new public service (Denhardt & Denhardt, 2000; 2003; Kumar, 2019) which builds on the idea of dialog between public managers and citizens in a spirit of democratic participation. This represents a considerable change from the previous approaches. Seven principles are delineated (Denhardt & Denhardt, 2003):

1. Serve citizens, not customers
2. Seek the public interest
3. Value citizenship over entrepreneurship
4. Think strategically, act democratically
5. Recognize that accountability is not simple
6. Serve rather than steer
7. Value people, not just productivity

They state that (Denhardt & Denhardt, 2002) "We argue here that the better contrast is with what we call the "New Public Service," a movement built on work in democratic citizenship, community and civil society, and organizational humanism and discourse theory" They go on to say " We suggest seven principles of the New Public Service, most notably that the primary role of the public servant is to help citizens articulate and meet their shared interests rather than to attempt to control or steer society". This approach takes public managers in a new direction and one that supports a more extensive view of public involvement.

Considering these three approaches, we can see a progression of support for robust public involvement. Figure one presents this relationship.

Table 1. Public involvement in public management theory

Old Public Administration > New Public Management > New Public Service

These three theories demonstrate an evolution from traditional public administration theory, which was a semi closed system doing the bidding of the political system, to a system that managed public service is semi isolation from public participation to a system that promotes democratic dialog in policy making. We are now at a point where the situation is about to change again and we begin to see the outlines of a new model.

Theories of practice should be seen in the context of the times when they were created. Public administration emerged in the early days of the industrial revolution along with a number of newer professional groups. Early management theory such as Taylorism (Taylor, 1924) was a response to the application of early knowledge to running public organizations. Since the 1970s (Porat, 1977) were have been dealing with an emerging information economy. While the Implications of this development are legion, three signature issues emerge for our discussion here—the growing sophistication of technology, the explosion of knowledge and the growing availability of data. This coupled with a changing social order and changing expectations for the way that things are managed insures that change will occur.

Most of the older public administration theory was developed for an industrial economy with lower levels of technology and different expectations and organizational models. As societal change occurs, the underpinnings of practice theories change. The New Public Management might be thought of as an early attempt to capitalize on the technology changes and societal progress that occurred as the information economy first emerged in the 1970s and 1980s. The New Public Service approach was formulated later on in the process. The youngest of these approaches is twenty years old. Many of the things that we take for granted today were not available at the turn of the millennium. While there was some effort to build smart cities decades ago, the effort was nothing like what we see today (Meijer & Bolívar, 2016; Meijer, Gil-Garcia & Bolívar, 2016). It is worth wondering if the fit between what is considered leading edge theory and the societal supports that it requires are still adequate.

Smart Cities, Open Data, Cocreation and the Changing Face of Public Management

In the past few decades, a number of forces have come together that challenge the insights of previous theory and provide an opportunity for a reformulation. These include the growth of smart cities, the emergence of open government, particularly the increasing availability of open data, the development of Web 2.0 and eventually Web 3.0 and the flowering of self-organizing systems and collaborative intelligence. Figure one presents this relationship:

Figure 1. Changes in the public administration ecosystem

Smart cities is an international movement incorporating technology to revolutionize the governance process and meet the needs of the rapidly growing urban population (David, Justice & McNutt, 2017; David & McNutt, 2019). This change the situation that public managers find themselves in (see Meijer, & Bolívar. 2016 for an excellent overview.). Leading smart cities is more complex than leading technology alone. This means that models of public administration must change to accommodate these new realities.

A related trend is the growth of transparency and open government (Dawes & Helbig,, 2010, August; Nam, 2012; Harrison, Guerrero, Burke, et, al, 2012). A key component of this trend is the provision of open data, which directly supports a number of smart city processes. This also create a number of complications for traditional models of public administration.

This began in the late 1990s with the emergence of Web 2.0 or social media (O'Reilly, 2005; 2007). and the creation of applications that facilitated user generated content and the pooling of collective ideas and intelligence (see McNutt, Guo, Goldkind & An, 2019; Budhathoki & Haythornthwaite, 2013). These include social networking sites, wikis, technology for knowledge management and so forth. There are also applications developed to catalog and access citizen expertise (Noveck, 2015).

The emergence of self-generated or self-organizing organizations (and theory to support that process) also tugged at the envelope (Lee & Edmondson, 2017). The idea that you could create organizations without formal organizing processes made traditional theory take notice. Leaderless organizations (Brainard, Boland & McNutt, 2018) also emerged. All of this challenged traditional ideas about organization and management theory in public affairs.

The emergence of massive amounts of data and its importance to the economy is difficult to ignore. This data is a driver of management in the commercial sector.

Finally, the move toward volunteering online has entered the arena, supporting the other two emergent forces. People contribute online to many efforts, some political and others not. One example is Wikipedia which is written by an army of online volunteers. Clay Shirky (2010) refers to this as cognitive surplus. There are also a substantial number of online social movements which vie for the opportunity to affect public policy.

All of this exists against a reduction in public support for government. The Pew Research Center (2019) found that:

Public trust in the government remains near historic lows. Only 17% of Americans today say they can trust the government in Washington to do what is right "just about always" (3%) or "most of the time" (14%).

After decades of devolution and privatization, reductions in the size of government now limits the capacity of government in many critical areas and the resources for government innovation are often just not there.

In the information technology area, governments are often far behind their corporate contemporaries. Tight budgets and constraining policies hamstring IT managers.

These forces have created a fertile ground for certain types of experimentation. While these movements are not earth shattering, in an of themselves, they do point to the possibility of future directions in citizen participation theory and the way that government can deal with citizens and other sectors. This will necessitate some serious rethinking of public administration theories.

New Approaches to Technology and Citizen Involvement

A number of innovations have begun to emerge against this backdrop that offer innovation on the periphery of government. Two of these movements are Civic Technology and Data4Good. These are efforts that make use of recent changes in technology and society. They may fall within the new public service or they might provide some of the beginning push for evolved theories of public administration.

Both approaches make use of highly skilled volunteers. While volunteers have long been used by government and the nonprofit sector, the assumption always was that they were low skilled individuals who could perform simple tasks. The volunteers in these two efforts are at or above the skill level of their counterparts in government. This means that managing them will require some different sills and a different orientation.

These are also movements that span the boundaries of the sectors. The commercial sector is highly involved here, as are nonprofit organizations and individual citizens. This is active involvement rather than coordination. It might be fairer to say that the boundaries are being blurred or even eliminated by these efforts. At the very least, much of the activity happens outside government.

Civic Technology

Civic technology is a technologically enhanced budding force in the relationship between government and communities. It is a worldwide movement that has interfaced nicely with the movement toward smart cities and a variety of other reform movements (McNutt, 2018; David, McNutt & Justice 2018; McNutt, Justice, Melitski, Ahn, Siddiqui, Carter & Kline, 2016; Living Cities, 2012; Gilman, 2016; Gordon & Mihailidis, 2016; Newsome, 2013; Schrock, 2016; 2018; Stephens, 2017; Suri, 2013). The emergence of civic technology, fueled by the movement toward open government and the ready provision of open civic data, could represent a sea change in the relationship between government and the communities that they serve.

Civic technology is defined by Living Cities (2012) as:

Civic technology is the use of digital technologies and social media for service provision, civic engagement, and data analysis – has the potential to transform cities and the lives of their low income residents. (Living Cities. 2012, 3).

Civic technology evolved outside of the government institution. While some contemporary civic technology efforts are cooperative with government, that isn't always the case. As the civic technology movement unfolds, it might very well be a force that challenges the domain of government in some quarters. In cooperation with a number of related online movements (see McNutt, 2018), civic technology can remake local governments in many profound ways. On balance, civic technology might continue to enhance the efforts of traditional government. The movement continues to evolve and how it might solidify is open to speculation.

In part, civic technology was a response to technology limitations in government. Government technology has often lagged behind the commercial sector in its use of technology. Many government organizations cannot afford the technology talent or equipment needed to implement the efforts their citizens need. Efforts like Code for America (www.codeforamerica.org) have addressed these technology skill deficits to some extent. As it has evolved, Civic technology has become far more than that.

The Mechanics of Civic Technology: Civic technology can be thought of as having three interrelated components: Open Civic Data, Civic technologies and Civic technology practices (McNutt, Justice, Melitski, Ahn, Siddiqui, Carter & Kline, 2016). These parts work together and create a common set of interventions. Again, there is often considerable variation in how this occurs in individual communities. Local conditions and the local talent pool influence how the elements are arranged.

The foundation of civic technology is open civic data. All of the other aspects of civic technology use data. This includes administrative data, sensor data, data from systems such as 311 systems and a huge variety of other data. The largest body of open civic data comes from different open government efforts (Lathrop & Ruma, 2010), which has expanded in the last two decades. Other data is from nonprofits, commercial organizations and citizen collected data. More and more, commercial organizations are sharing data with government and with some nonprofits. For example, ride sharing companies like Uber share their data with transportation planners. This data can also be shared with other organizations through data collaboratives, which are organized efforts to facilitate the sharing of data (Verhulst & Sangokoya, 2015; Susha, Janssen, Verhulst & Pardo, 2017, June). Citizens groups also collect data. Citizen science organizations (Kullenberg & Kasperowski, 2016), for example, work with researchers in more formal settings with volunteers doing much of the data collection. This is a substantial part of scientific data collection.

Civic technology pairs open civic data with technology and a series of social innovations like crowdsourcing and hackathons (Powell, 2016; McNutt & Justice,

2016; Irani, 2015; Hou & Lampe, 2017, June) and peer networks (such as Code for America Brigades). This can create a peripheral or possibly parallel structure.

The technological arsenal of civic technology is a mix of technology created for other reasons and repurposed and technology that was created expressly for civic technology needs. An example of the former is My Society's Fix My Street system (King & Brown, 2007). This is a website that lets drivers report road hazards. The site provides the data to public works and monitors if the problem is addressed. If the problem is not addressed, the site publicizes that fact. Almost any technology could potentially fall into the second category. Organizations typically use a wide range of social media/web 2.0 applications, databases and other systems.

Finally, there are civic technology practices. Most of these are based on well-known processes that have been adapted for civic technology. Crowdsourcing is a time honored practice and coproduction has been part of the public policy literature for many decades. Some of the major civic technology practices are peer groups (such as Code for America's Brigades), Hackathons and other contests and crowdfunding/crowdsourcing. Code for America's Brigades are groups of locally based volunteer technologists who work on a range of projects in their local communities. Hackathons (McNutt & Justice, 2016; Irani, 2015) are event where participants work on projects that are defined by community organizations. The National Day of Civic Hacking is a Code for America effort that encourages hackathons in different communities on a single day. These three realms work together to make civic technology fundamentally different from traditional government computing and e-government. Active citizen involvement (or even citizen control) is part of the DNA of civic technology. This would take us to the top of Arnstein's (1969) classic ladder of citizen participation.

Civic technology is supported and promoted by a number of organizations. The most prominent is Code for America (www.codeforamerica.org), a nonprofit that organizes local brigades and sponsors Code for America Fellowships and also sponsors the National Day of Civic Hacking. Their international effort, Code for All (www.codeforall.org/), promotes civic technology globally. The Knight Foundation (Patel, Sotsky, Gourley & Houghton, 2013) and Living Cities (2012) were early supporters of civic technology. My Society (www.mysociety.org/) is a major sponsor in the United Kingdom. The Ford Foundation's Technology for Good effort will add substantially to the growth of these efforts. Academic homes include MITs Center for Civic Media, the Engagement Lab at Emerson College and a number of other efforts.

Civic technology runs the gamut from providing technology for local government to creating parallel institutions that exist alongside of government. Much of the effort is performed by volunteer technologists who are employed by other organizations. At the very least, it is a boundary spanning situation. Civic technology can also represent parallel government structures that might signify a new stage in public administration and public policy.

There are a number of efforts that are similar to civic technology in some way and work well with the concept. Participatory Budgeting is one such effort. In this system, residents are allowed to vote on various municipal expenditures and those decisions are binding. This is a worldwide effort with examples in many places.

A related, although separate, approach is data4good. Like civic technology, it uses data, highly skilled volunteer involvement and the incorporation of the nonprofit and commercial sector. While civic technology is focused on technology development, data4good is more focused on analytics and data science. There are definite overlaps between the two forces and their interests converge.

Data for Good

Data Science is an emerging force in society that aims to use massive amounts of data combined with sophisticated analytical techniques and a range of technologies to solve complex problems that were once beyond our ability to address (Kelleher & Tierney, 2018). Data4good supports the use of this technology by governments and nonprofits.

Combining technology that can be used to process large quantities of data with analytical capacity that can deal with huge data sets, this developing area is growing quickly and those trained in data science are in high demand. The core disciplines include statistics, computer science and mathematics. Tools like machine learning and artificial intelligence are key components of data science. Data science also puts a premium on content or domain expertise and the idea of "ground truth".

The growth in the volume of data in recent years has been enormous. Administrative data has been joined by output from social media and sensor data created by the internet of Things. Data is now collected in a series of unstructured formats. This data holds the promise of creating new knowledge and facilitating the development of evidence based policies.

Analytical capacity has also improved greatly. New approaches can handle the greater volume of data and make more complex analysis possible. Artificial intelligence and machine learning facilitate human analysis.

Data science has been a major asset to financial services firms, engineering efforts, marketing, pharmaceutical research and a variety of other fields. Political organization have also found data scientists useful (for example consider the role played by Cambridge Analytica in the 2016 US election). Those trained in data science can command high salaries. This often puts such employees out of reach for many public and nonprofit organizations.

Data4Good is a movement that brings data science expertise to the solution of pressing societal issues such as poverty, disease, human trafficking and slavery, climate change and so forth (Howson, Beyer, Idoine, Jones, 2018; Bull, Slavitt &

Lipstein, 2016; Pfaff, 2015). The huge growth in available data from administrative sources, social media and sensor and tracking data makes addressing many social issues possible, given the ability to manage and analyze that data. The emerging field of data science offers immense capacity to address these challenges that government and nonprofits face. Sadly, few of these organizations can afford the cost of data science practitioners. This means that potential remains potential.

Data for Good is an attempt to remedy that situation. Organizations that use data for good strategies use volunteer data scientists to help in their work. Many, if not most of these workers have jobs in other sectors such as financial analytics, manufacturing or technology. The ability to work on projects that are intellectually or personally fulfilling is often an important motivation for involvement in data4good projects. Consequently, firms that employ data scientists are often willing to let their employees work on projects during work time or provide stipends for them to work in another setting (Howson, Beyer, Idoine & Jones, 2018; Pfaff, 2015). This builds the employer-employee relationship and contributes to employee commitment. It also gives staff the opportunity to master new methods.

Table 2. The data for good ecosystem

Data For Good Ecosystem			
Employers	Placement Organizations	Data Providers	Domain Organizations

Making data for good strategies work entails a constellation of organizations that can provide different resources. Employer participation is one factor. Some organizations provide opportunities with external efforts while others mount their own campaigns. There are also data scientists and other professionals who volunteer their time without employer involvement. The second group of organizations are those that match volunteers with data science opportunities. These organizations broker assignments and support volunteers. This requires substantial expertise and a wide network of relationships. Data Kind (www.datakind.org) is an excellent example and probably the preeminent organization working in this space. Third, there are providers of data. Some of the data providers are government organizations while others are nonprofits or commercial organizations. Some of these organizations are boundary organizations that connect the research community with organizations. Data Collaboratives (Verhulst & Sangokoya, 2015) represent an emerging form of arrangement that brokers corporate data released as data philanthropy (McKeever, Greene, MacDonald, Tatian, & Jones, 2018) and through other efforts. Finally, there are the domain nonprofits who are working on the issue. In some cases, this could be a government organization.

In addition to these formal efforts, there are more short term solutions. Similar to Hackathons are *Data Dives*. Pfaff (2015) explains that "Data Dives are weekend-long, marathon-style events where dozens of volunteers rally together to help 3-4 social change organizations do initial data analysis, exploration, and prototyping. These events are free for organizations, open to volunteers of all skill levels and take place around the world."

Bloomberg, the Rockefeller Foundation and MasterCard have invested heavily in these activities, as have SAS (Statistical analysis) and a number of technology companies. Professional associations, such as Association for Computing Machinery (ACM) are also involved. There are conferences and meeting on data4good throughout the world. Many of the organization that were involved in civic technology are also part of this movement

Data4Good uses some of the same processes as civic technology. It is also related to Data Journalism and Data Activism and the Data Justice movement. Data Journalism is the application of data science to reporting and investigative journalism. It is a specialty in the journalism profession. Data Activism (Puussaar, Johnson, Montague, James & Wright, 2018) is a branch of social movement activity that makes use of data and data science. The data justice movement looks at the larger issues involved in data and its impact on society.

Civic Technology and Data for Good Compared

Civic technology and Data4Good are complementary techniques that look at the change process in different ways. There are definite similarities—the importance of data and the use of highly skilled volunteers. The central role of technology, but generally different technologies, is critical. There are also a number of important differences. Table 1 compares the two practices at a very general level.

Civic technology tends to operate at a community or regional level while Data for Good is usually more of a policy level intervention. There are, however, many exceptions. The growth of national civic technology networks will continue to blur that distinction and there are a number of activities that can far into either camp.

The differences between either of these efforts and traditional e-government are substantial. How those differences are accommodated (in some cities there are truly wonderful partnerships) will determine if these efforts can make the needed contributions.

Table 3. A comparison of civic technology and Data4Good

Aspect	Civic Technology	Data for Good
Data	Open Civic Data/some involvement of community data from partners	Typically, large datasets, Data from Data Philanthropy Efforts and other sources
Technology	A range of technologies depending on the issues	Typically, analytics, GIS, Artificial Intelligence & Machine Learning
Community Involvement	Multisector collaboration	Could be limited to the nonprofit or group making the request
Volunteers	Highly Skilled Technologists and other Volunteers	Data Scientists and other skilled volunteers
Change Process	Creation of technology solutions in communities. This is general consensus based but might involve pressure tactics.	Creation of Information to power other change processes. Techniques like data storytelling and data activism are also involved.

Implications

These two emerging approaches, in their current versions, will have minimal direct impacts on how e-government is conducted. As a proof of concept, however, the potential impact is substantial. The current programs point the way towered some interesting new directions. These organizations can create new efforts that substantially change the future of the field.

Most of these efforts are small. While they can make important conceptual contributions to public policy and administration, running the entire enterprise in this way would not be possible. It is difficult to conceptualize a future where government is a function of volunteer labor and the slack resources of other sectors. In addition, politics is part and parcel of the conduct of government and requires a great deal of effort, particularly when issues are controversial. Making major change is always difficult. Small organizations are not always up for a long term battle that requires considerable resources.

The principle contributions are (1) material contributions to specific issues in communities and (2) the creation of prototypes that can be expanded upon in other communities. Given the scope of current efforts, the long term benefit might be in the prototyping, although the ability to replicate the models could have the greatest long term benefit. These approaches auger well with smart city models that are being developed and could address the citizen involvement that some smart city models ignore.

The process by which these concepts become innovations is not as straightforward as it might be (see Rogers, 2003), but there are new wrinkles that can make it more possible. Policy Labs and Innovation Labs are possible solutions (Williamson, 2015).

These are movements that could coordinate with existing social movement to build broader social change. These are efforts that would appeal to many movements, on the left and right, that are interested in reforming government. One also wonders how the growing familiarity of highly skilled and highly educated volunteers with the problems of government will impact attitudes toward government. These could be efforts that could address the trust in government issue.

The growth of alternatives, such as civic technology and Data4Good, represents an interesting trend. Whether they constitute a harbinger of things to come is still unclear.

What is clear is that public administration practice theory must evolve to meet new developments. Progress on the ground suggest that concurrent progress in theory is needed.

REFERENCES

Andrisani, P., Hakim, S., & Savas, E. S. (2002). *The New Public Management: Lessons from Innovating Governors and Mayors*. Boston: Kluwer Academic Publishers. doi:10.1007/978-1-4615-1109-0

Arnstein, S. R. (1969). A ladder of citizen participation. *Journal of the American Institute of Planners*, *35*(4), 216–224. doi:10.1080/01944366908977225

Baraniuk, C. (2013). The civic hackers reshaping your government. *New Scientist*, *218*(2923), 36–39. doi:10.1016/S0262-4079(13)61625-5

Brainard, L., Boland, K., & McNutt, J. G. (2018). The Advent of Technology Enhanced Leaderless Transnational Social Movement Organizations: Implications for Transnational Advocacy. (185-). In J. G. McNutt (Ed.), *Technology, Activism and Social Justice in a Digital Age*. London: Oxford University Press.

Brainard, L., & McNutt, J. G. (2010). Virtual Government-Citizen Relations: Old Public Administration, New Public Management or New Public Service? *Administration & Society*, *42*(7), 836–858. doi:10.1177/0095399710386308

David, N., Justice, J. B., & McNutt, J. G. (2015). Smart Cities are Transparent Cities: The Role of Fiscal Transparency in Smart City Governance. In P. R. B. Manuel (Ed.), *Transforming City Governments for successful Smart Cities. Empirical Experiences*. Berlin: Springer. doi:10.1007/978-3-319-03167-5_5

David, N., Justice, J. B., & McNutt, J. G. (2018). Smart Cities, Transparency, Civic Technology and Reinventing Government. In Smart Technologies for Smart Governments. Transparency, Efficiency and Organizational Issues. Berlin: Springer.

Dawes, S. S., & Helbig, N. (2010, August). Information strategies for open government: Challenges and prospects for deriving public value from government transparency. In *International Conference on Electronic Government* (pp. 50-60). Springer. 10.1007/978-3-642-14799-9_5

Dawes, S. S., Vidiasova, L., & Parkhimovich, O. (2016). Planning and designing open government data programs: An ecosystem approach. *Government Information Quarterly*, *33*(1), 15–27. doi:10.1016/j.giq.2016.01.003

Denhardt, J., & Denhardt, R. (2000). The New Public Service: Serving Rather than Steering. *Public Administration Review*, *60*(6), 549–559. doi:10.1111/0033-3352.00117

Denhardt, J. V., & Denhardt, R. B. (2003). *The new public service: serving not steering*. Armonk, NY: M.E. Sharpe.

Denhardt, R. (1999). *Public Administration: An Action Orientation* (3rd ed.). Fort Worth, TX: Harcourt Brace.

Desouza, K. C., & Bhagwatwar, A. (2012). Citizen apps to solve complex urban problems. *Journal of Urban Technology*, *19*(3), 107–136. doi:10.1080/10630732.2012.673056

Desouza, K. C., & Bhagwatwar, A. (2014). Technology-Enabled Participatory Platforms for Civic Engagement: The Case of US Cities. *Journal of Urban Technology*, *21*(4), 25–50. doi:10.1080/10630732.2014.954898

Gilman, H. R. (2016). *Participatory budgeting and civic tech: The revival of citizen engagement*. Washington, DC: Georgetown University Press.

Gordon, E., & Mihailidis, P. (Eds.). (2016). *Civic media: Technology, design, practice*. Cambridge: MIT Press. doi:10.7551/mitpress/9970.001.0001

Gruening, G. (2001). Origin and theoretical basis of New Public Management. *International Public Management Journal*, *4*(1), 1–25. doi:10.1016/S1096-7494(01)00041-1

Harrison, T. M., Guerrero, S., Burke, G. B., Cook, M., Cresswell, A., Helbig, N., & Pardo, T. (2012). Open government and e-government: Democratic challenges from a public value perspective. *Information Polity*, *17*(2), 83–97. doi:10.3233/IP-2012-0269

Harrison, T. M., Pardo, T. A., & Cook, M. (2012). Creating open government ecosystems: A research and development agenda. *Future Internet*, *4*(4), 900–928. doi:10.3390/fi4040900

Hébert, M. K. (2014). Come Hack with Me: Adapting Anthropological Training to Work in Civic Innovation. *Practical Anthropology*, *36*(2), 32–36. doi:10.17730/praa.36.2.405j1uvvn8584768

Hood, C. (1991). A public management for all seasons? *Public Administration*, *69*(1), 3–19. doi:10.1111/j.1467-9299.1991.tb00779.x

Hou, Y., & Lampe, C. (2017, June). Sustainable hacking: characteristics of the design and adoption of civic hacking projects. In *Proceedings of the 8th International Conference on Communities and Technologies* (pp. 125-134). New York: ACM. 10.1145/3083671.3083706

Howson, C., Beyer, M. A., Idoine, C. J., & Jones, L. C. (2018). *How to Use Data for Good to Impact Society*. New York: Gartner. https://www.gartner.com/doc/3880666/use-data-good-impact-society

Irani, L. (2015). Hackathons and the Making of Entrepreneurial Citizenship. *Science, Technology & Human Values*, *40*(5), 799–824. doi:10.1177/0162243915578486

Justice, J., McNutt, J. G., Melitski, J., Ahn, M., David, N., Siddiqui, S., & Ronquillo, J. C. (2018). The Civic Technology Movement: Implications for Nonprofit Theory and Practice. (89). In J. G. McNutt (Ed.), *Technology, Activism and Social Justice in a Digital Age*. London: Oxford University Press.

Kelleher, J. D., & Tierney, B. (2018). *Data science*. Cambridge, MA: MIT Press. doi:10.7551/mitpress/11140.001.0001

King, S. F., & Brown, P. (2007, December). Fix my street or else: using the internet to voice local public service concerns. In *Proceedings of the 1st international conference on Theory and practice of electronic governance*. (pp. 72-80). New York: ACM. 10.1145/1328057.1328076

Kullenberg, C., & Kasperowski, D. (2016). What Is Citizen Science? – A Scientometric Meta-Analysis. *PLoS One*, *11*(1), e0147152. doi:10.1371/journal.pone.0147152 PMID:26766577

Kumar, A. (2019). Citizen-centric model of governmental entrepreneurship: Transforming public service management for the empowerment of marginalized women. *Transforming Government: People. Process and Policy*, *13*(1), 62–75.

Lathrop, D., & Ruma, L. (Eds.). (2010). Open Government: Collaboration, Transparency, and Participation in Practice. Sevastopol, CA: O'Reilly.

Lee, M. Y., & Edmondson, A. C. (2017). Self-managing organizations: Exploring the limits of less-hierarchical organizing. *Research in Organizational Behavior*, *37*, 35–58. doi:10.1016/j.riob.2017.10.002

McKeever, B., Greene, S., MacDonald, G., Tatian, P., & Jones, D. (2018). *Data Philanthropy: Unlocking the Power of Private Data for Public Good*. Washington, DC: Urban Institute.

McNutt, J. G. (2018) (ed). Technology, Activism and Social Justice in a Digital Age. London: Oxford University Press.

McNutt, J. G., Brainard, L., Zeng, Y., & Kovacic, P. (2016). Information and Technology In and For Associations and Volunteering. In Palgrave Handbook of Volunteering and Nonprofit Associations. Basingstoke, UK: Palgrave Macmillan.

McNutt, J. G., Guo, C., Goldkind, L., & An, S. (2018). Technology in Nonprofit organizations and voluntary action. *Voluntaristics Review*, *3*(1), 1–63. doi:10.1163/24054933-12340020

McNutt, J. G., & Justice, J. B. (2016, September). *Predicting civic hackathons in local communities: Perspectives from social capital and creative class theory*. Presented at the *ISTR Conference*, Stockholm, Sweden.

McNutt, J. G., Justice, J. B., Melitski, M. J., Ahn, M. J., Siddiqui, S., Carter, D. T., & Kline, A. D. (2016). The diffusion of civic technology and open government in the United States. *Information Polity*, *21*(2), 153–170. doi:10.3233/IP-160385

Meijer, A., & Bolívar, M. P. R. (2016). Governing the smart city: A review of the literature on smart urban governance. *International Review of Administrative Sciences*, *82*(2), 392–408. doi:10.1177/0020852314564308

Meijer, A., & Bolívar, M. P. R. (2016). Governing the smart city: a review of the literature on smart urban governance. *International Review of Administrative Sciences, 82*(2), 392-408.

Meijer, A. J., Gil-Garcia, J. R., & Bolívar, M. P. R. (2016). Smart city research: Contextual conditions, governance models, and public value assessment. *Social Science Computer Review*, *34*(6), 647–656. doi:10.1177/0894439315618890

Nam, T. (2012). Citizens' attitudes toward open government and government 2.0. *International Review of Administrative Sciences*, *78*(2), 346–368. doi:10.1177/0020852312438783

Newsome, G. (2013). *Citizenville*. New York: The Penguin Press.

Noveck, B. S. (2015). *Smart citizens, smarter state: The technologies of expertise and the future of governing*. Cambridge: Harvard University Press. doi:10.4159/9780674915435

O'Reilly, T. (2005). *Web 2.0: compact definition.* Message posted to http://radar. oreilly. com/archives/2005/10/web_20_compact_definition. html

O'Reilly, T. (2007). What is Web 2.0: Design patterns and business models for the next generation of software. *Communications & Stratégies, 65*(1), 17–37.

Patel, M., Sotsky, J., Gourley, S., & Houghton, D. (2013). *The emergence of civic tech: Investments in a growing field*. The Knight Foundation.

Pew Research Center. (2019, April 11). *Public Trust in Government: 1958-2019.* https://www.people-press.org/2019/04/11/public-trust-in-government-1958-2019/

Pfaff, T. (2015). *The Definitive Guide to doing Data Science for Social Good.* https://www.kdnuggets.com/2015/07/guide-data-science-good.html

Powell, A. (2016). Hacking in the public interest: Authority, legitimacy, means, and ends. *New Media & Society, 18*(4), 600–616. doi:10.1177/1461444816629470

Puussaar, A., Johnson, I. G., Montague, K., James, P., & Wright, P. (2018). Making open data work for civic advocacy. *Proceedings of the ACM on Human-Computer Interaction, 2*(CSCW), 143. 10.1145/3274412

Reiter, R., & Klenk, T. (2019). The manifold meanings of 'post-New Public Management'–a systematic literature review. *International Review of Administrative Sciences, 85*(1), 11–27. doi:10.1177/0020852318759736

Rogers, E. M. (2003). *The Diffusion of innovation* (5th ed.). New York: Free Press.

Rumbul, R. (2015). *Who Benefits From Civic Technology? Demographic and public attitudes research into the users of civic technology*. London: mySociety.

Schlozman, K. L., Brady, H., & Verba, S. (2018). *Unequal and unrepresented: Political inequality and the people's voice in the new guided age*. Princeton, NJ: Princeton University Press.

Schrock, A., & Shaffer, G. (2017). Data ideologies of an interested public: A study of grassroots open government data intermediaries. *Big Data & Society, 4*(1), 2053951717690750. doi:10.1177/2053951717690750

Schrock, A. R. (2016). Civic hacking as data activism and advocacy: A history from publicity to open government data. *New Media & Society, 18*(4), 581–599. doi:10.1177/1461444816629469

Shirky, C. (2010). *Cognitive surplus: Creativity and generosity in a connected age.* New York: Penguin.

Stepasiuk, T. (2014). Civic hacking: A Motivational Perspective. *New Visions in Public Affairs., 6*, 21–30.

Stephens, J. (2017). *Civic Technology: Open Data and Citizen Volunteers as a Resource for North Carolina Local Governments.* Chapel Hill, NC: School of Government, The University of North Carolina at Chapel Hill.

Suri, M. V. (2013). *From Crowdsourcing Potholes to Community Policing: Applying Interoperability Theory to Analyze the Expansion of "Open311".* Berkman Center for Internet & Society at Harvard University.

Susha, I., Grönlund, Å., & Van Tulder, R. (2019). Data driven social partnerships: Exploring an emergent trend in search of research challenges and questions. *Government Information Quarterly, 36*(1), 112–128. doi:10.1016/j.giq.2018.11.002

Susha, I., Janssen, M., & Verhulst, S. (2017). Data collaboratives as "bazaars"? A review of coordination problems and mechanisms to match demand for data with supply. *Transforming Government: People. Process and Policy, 11*(1), 157–172.

Susha, I., Janssen, M., Verhulst, S., & Pardo, T. (2017, June). Data collaboratives: How to create value from data for public problem solving? In *Proceedings of the 18th Annual International Conference on Digital Government Research* (pp. 604-606). New York: ACM. 10.1145/3085228.3085309

Taylor, F. (1923). *Scientific Management.* New York: Harper and Row.

Verba, S., Schlozman, K., & Brady, H. (1995). Voice and Equality: Civic Voluntarism. In *American Politics.* Cambridge: Harvard University Press.

Verhulst, S., & Sangokoya, D. (2015). *Data collaboratives: Exchanging data to improve people's lives.* https://medium. com/@ sverhulst/data-collaboratives-exchanging-data-to-improvepeople-s-lives-d0fcfc1bdd9a

Williamson, B. (2015). Governing methods: Policy innovation labs, design and data science in the digital governance of education. *Journal of Educational Administration and History, 47*(3), 251–271. doi:10.1080/00220620.2015.1038693

Chapter 7

Determinants of the Citizen Engagement Level of Mayors on Twitter:
The Case of Turkey

İbrahim Hatipoğlu
 https://orcid.org/0000-0002-4561-9160
Bursa Uludağ University, Turkey

Mehmet Zahid Sobaci
 https://orcid.org/0000-0003-2625-145X
Bursa Uludağ University, Turkey

Mehmet Fürkan Korkmaz
 https://orcid.org/0000-0001-6141-5777
Bursa Uludağ University, Turkey

ABSTRACT

Today, politicians like other political actors use social media to interact with their audiences. In the relevant literature, studies on the use of social media by politicians focus more on how politicians use social media for political communication during the election periods and its impact on the election results. Furthermore, these studies mainly focus on national politicians. Few studies focus on the use of social media during a non-election period by the local politicians, and these studies analyse the purpose of using social media. Therefore, in the relevant literature, there is a need for empirical studies to measure the citizen engagement level of local politicians during the non-election period and analyse its determinants beyond the purpose

DOI: 10.4018/978-1-7998-1526-6.ch007

of using social media. In this context, this study aims to analyse the relationship between some factors and the level of citizen engagement of the mayors on Twitter in Turkey. The findings of the analysis show that there is a relationship between the status of municipalities and the engagement level of mayors.

INTRODUCTION

Social media has become a part of people's daily lives in the last decades by dint of the proliferation of low-priced Internet devices. The rise of social media platforms has changed people's communication, shopping and entertainment styles. Advantages of social media were first discovered by the private sector like other technological developments, especially for brand marketing and customer-focused management. Afterwards, social media drew the attention of political actors (Sobaci, Hatipoglu, & Korkmaz, 2018). The political actors such as political parties, politicians and activists have begun to use social media tools to organise, mobilise and engage to their audiences (Larsson & Moe, 2014; Vergeer, Hermans, & Sams, 2013; Kalnes, 2009; Stranberg, 2013; Eltantawy & Wiest, 2011). In this context, the use of social media by politicians is not a new phenomenon. However, today, social media are more intensively used by national as well as local politicians (Sobaci & Karkin, 2013).

Social media tools such as Twitter are favourable for online political marketing at the local level. These tools fulfil the needs of local politicians for "a personal, direct, interactive, and speed style" of communication with citizens (Criado, Martínez-Fuentes, & Silván, 2012). Social media are direct and probably the cheapest way for a political campaign during the election period. Therefore, most of the sub-national politicians, as well as national politicians, use social media in their election campaigns (Triantafillidou, Lappas, Kleftodimos, & Yannas, 2018; Larsson, 2018; Welp, Capra, & Freidenberg, 2018). Moreover, the local politicians, especially mayors, use social media tool for engaging the citizens and promoting themselves during the non-election periods.

In the relevant literature, studies on the use of social media by politicians have focused on how politicians use social media for political communication during the election periods and its impact on the election results (Hansen & Kosiara-Pedersen, 2014; Strandberg, 2013; Carlson & Stranberg, 2008; Ozdeşim İkiz, Sobaci, Yavuz, & Karkin, 2014; Welp et al., 2018; Lev-On, 2018). Furthermore, these studies have mainly focused on national politicians (Williams & Gulati, 2013; Hansen & Kosiara-Pedersen, 2014; Strandberg, 2013; Carlson & Stranberg, 2008). Few studies, in contrast, have focused on the use of social media during a non-election period by the local politicians, and these studies have analysed the purpose of the using social media (Sobaci & Karkin, 2013; Vučković & Bebić, 2013). Therefore, in the relevant literature, there is a need for empirical studies to measure the citizen engagement level of local politicians during the non-election period and to analyse its determinants beyond the purpose of using social media.

In this context, this study aims to analyse the relationship between factors, that are personal traits of mayors (age, gender, education), characteristics of municipalities (region and status), and political context (terms of mayor, and mayors' political party), and the level of citizen engagement of the mayors on Twitter in Turkey. This study is organised into five sections. The second section presents the literature review of relationship citizen engagement and social media. Also, literature about the determinants of citizen engagement is reviewed in this section. The third section reveals the methodology of the study, including data collection, measuring the engagement level and statistical methods. The fourth section presents the findings of the study. The study concludes with the discussion of the results.

LITERATURE REVIEW

Citizen Engagement and Social Media

The concept of public engagement is broadly defined as the involvement of citizens in social issues (Rowe & Frewer, 2005). Public engagement is different from traditional interaction between politicians and citizens because it is based on a two-way flow of information (Sheedy, 2008). Developments in information and communication technologies in the last two decades have contributed significantly to the two-way flow of information. Therefore, public engagement has become one of the most trend-topic during the last two decades. Especially with the developments of Web 2.0 technologies, there has been an increase in the number of academic studies examining public engagement (Skoric, Zhu, Goh, & Pang, 2016).

Web 2.0 has brought a new dimension to the use of ICTs in politics as well as many different areas. When O'Reilly (2007) defined the concept of Web 2.0, he urged on its "user control" feature to create, design and develop the content and services to contradistinguish it from other ICTs development. Chu and Xu (2009, p.717) described Web 2.0 as "is of the user, by the user, and, more importantly, for the user". Since social media was developed based on this philosophy, its nature involves user-generated content and two-way communication (Sobaci et al., 2018). Therefore, it can be a powerful tool to promote citizen engagement (Criado, Sandoval-Almazan, & Gil-Garcia, 2013; Bode, 2012; Skoric et al., 2016).

Previous studies examined the relationship between social media and public engagement from different perspectives. The first group of studies analysed the impact of social media on citizen engagement. Zhang, Johnson, Seltzer, and Bichard (2010) examined the extent to which reliance on social networks such as Facebook, MySpace and Youtube have engaged citizens in political activities. They claimed that there is a positive relationship between social media use and citizen engagement. Valenzuela, Park, and Kee (2009) found that the use of Facebook among college students is positively associated with their engagement. According to Skoric et al.'s (2016) meta-analytic review, most of the researches have the same argument in the literature examining the impacts of social media on engagement.

The other group of researchers measured the level of engagements through social media by using metric sets. Most of these researches in this group focused on local governments (Bonsón, Ratkai, & Royo, 2016; Agostino & Arnaboldi, 2016; Triantafillidou, Lappas, Yannas, & Kleftodimos, 2015). In addition to these, there are a few studies focused on national level actors such as political parties (Sobacı & Hatipoğlu, 2016). The last group of studies tried to determine which factors influence the engagement level on social media. Bonsón, Perea, and Bednárová (2019) identified the relationship between engagement level and factors such as municipality size, audience, activity, media type and content type. Like the second group, most of the researches in this group focused on local governments (Bonsón, Royo, & Ratkai, 2015; Haro-de-Rosario, Sáez-Martín, & Carmen Caba-Pérez, 2018; Sobacı et al., 2018). Nevertheless, there is only one research focused on determinants of the engagement level of mayors on social media (Szmigiel-Rawska, Łukomska, & Tavares, 2018).

Determinants of Citizen Engagement

The main aim of this study is to determine which factors influence citizen engagement. These factors can be considered as three groups: the personal traits of mayors (gender, age, education level), characteristics of a municipality (status, and region), and political context (mayors' political party, term and political competition). Therefore, eight research questions were formulated to determine factors.

Firstly, we examined the relationship between gender and citizen engagement. Previous research showed that women prefer more participatory leadership type (Eagly & Johnson, 1990). Moreover, Tavares and Cruz (2017) argued that there is a relationship between gender and municipal transparency. In this context, our first research question is:

RQ1: Is there any relationship between the gender of mayors and the level of citizen engagement on Twitter?

The other important characteristic of mayors is age. In general, older people do not tend to adopt technological innovations (Morris & Vankatesh, 2000). Szmigiel-Rawska et al. (2018) found that younger mayor is more likely to successfully engage citizens via social media. To examine this relationship, the second research question is:

RQ2: Is there any relationship between age of mayors and the level of citizen engagement on Twitter?

Thirdly, we test whether the educational background influences the engagement level of mayors on Twitter. It is expected to that mayors who have higher education degree are successful for engaging citizens on social media. To analyse this phenomenon, our third research question is:

RQ3: Is there any relationship between the education level of mayors and citizen engagement on Twitter?

In addition to personal traits of mayors, some factors which can be categorized as characteristics of a municipality can influence the citizen engagement. The first factor as a characteristic of a municipality is municipal status. In Turkey, the municipal status is directly related to population of municipality. Previous empirical research (Bonsón, Royo, & Ratkai, 2017; Bonson et al., 2019) claimed that there is no significant relationship between municipality size and engagement level. To examine this phenomenon, our fourth research question is:

RQ4: Is there any relationship between municipal status (metropolitan or provincial) and level of citizen engagement on Twitter?

Due to the differences in levels of economic development among geographical region in Turkey, the engagement level of mayors may vary in a different region. To answer this question, the fifth research question is:

RQ5: Is there any relationship between the geographical region of municipality and level of citizen engagement on Twitter?

Political competition is likely to affect the use of social media by mayors. Previous research found that there is a positive relationship between political competition and the disclosure of information on municipal web pages (Gandía & Archidona, 2008; García & García-García, 2010). Moreover, Szmigiel-Rawska et al. (2018) pointed out that there is a positive relationship between political competition and engagement level of mayors on social media. In this context, the sixth research question is:

RQ6: Is there any relationship between political competition and level of citizen engagement on Twitter?

In the relevant literature previous research has analysed the influence of the ideology of the governing party on the engagement level (Haro-de-Rosario et al., 2018). This study argued that there is no significant relationship between political context and engagement level. Therefore the seventh research question is:

RQ7: Is there any relationship between political parties of mayors and level of citizen engagement on Twitter?

The last factor about the political context is terms of mayors. It is assumed that mayors' previous involvement contributes to their popularity and so it increases their engagement level. Szmigiel-Rawska et al. (2018) found that mayors term influence their engagement level. Therefore, the last research question is:

RQ8: Is there any relationship between the mayor's term and level of citizen engagement on Twitter?

METHODOLOGY

The empirical investigation of the study was conducted in five stages: i) identifying the official Twitter accounts of the mayors; ii) data collection about tweets and followers; iii) measuring the level of engagement using a metrics set; iv) data collection about mayors v) analysing the relationship between the characteristics of mayors and the level of citizen engagement. In this context, initially, the mayors' Twitter accounts were identified. For this, the search function of Google and Twitter was used. Moreover, the official web sites of the municipalities and Twitter accounts of the municipalities were used to avoid fake account.

In the second phase of the study, data regarding the tweets sent by the mayors during the investigation period were obtained using the package "rtweet" (Kearney, 2018) for the programming software R. The obtained data using the package "rtweet" include various pieces of information such as the number of friends and followers, the number of favourites and retweets on each post, the posting date and time. In this study, as it aims to measure the current status of the citizen engagement level of the mayors, the tweets from each mayor during 17 April-16 May 2019 was focused on. The reason for April 17 chosen as a starting day of the period is that all the mayors who include this study had received mayoral mandate before that day.

In the third stage of the study, the levels of engagement of mayors with the citizens were measured by using a metric set. In the relevant literature, there are various methods for measuring citizen engagement through social media. However, two main approaches dominate the literature. The first group studies analysed citizen engagement level using questionnaires (De Zúñiga, Copeland, & Bimber 2013; Warren, Sulaiman, & Jaafar, 2014; Paek, Hove, Jung, & Cole 2013). The second group studies propose a metrics set to analyse citizen engagement through social media without the use of questionnaires (Haro-de-Rosario et al., 2018; Bonsón et al., 2017; Bonsón & Ratkai, 2013; Agostino & Arnaboldi, 2016; Sobaci & Hatipoglu, 2017).

In this research, the level of engagement of mayors was measured by using a metric set under the second approach. It is based on the revised version of the metrics set proposed by Bonson et al. (2016, 2019). Due to the Twitter REST API restriction, the number of comments per tweet could not be gotten. Therefore, the number of comment per tweet was ignored for measuring engagement level. Table 1 illustrates the metric sets in detail.

Table 1. Metrics set for measuring engagement

Favourites	F1	Total favourites/total number of tweets	Average number of favourites per tweet
	F2	(F1/number of followers) x 1000	Average number of favourites per tweet per 1000 followers
Retweets	R1	Total retweets/total tweets	Average number of shares per post
	R2	(R1/number of followers) x 1000	Average number of retweets per tweet per 1000 followers
Engagement	**E**	**F2+R2**	**Engagement Index**

Source: (Bonson et al. 2016; 2019)

In the fourth stage of the study, information about the mayors including gender, age, education level, and political party, and municipal status, geographical region of a municipality, political competition and terms were collected. For this, biographies of the mayors on the municipalities' websites and the other online platforms such as the official website of mayors and website of Supreme Election Council were examined. Finally, the relationship between the characteristics of mayors and the level of citizen engagement were analysed.

Table 2. Variables and sources

Name	Description	Source
Gender	Gender of Mayor	Official Websites of Municipalities and Mayors, access May 2019
Age	Age of Mayor in 2019	Official Websites of Municipalities and Mayors, access May 2019
Education Background	The highest degree of school mayors have completed	Official Websites of Municipalities and Mayors, access May 2019
Term	The number of terms served by mayor	Official Websites of Municipalities and Mayors, access May 2019
Status of Municipality	Metropolitan and Provincial	Official Websites of Municipalities, access May 2019
Region of Municipality	The geographical region of municipality in Turkey	Official Websites of Turkish Geographical Society, access May 2019
Political Parties of Mayor	Political Party which the Mayor Belongs	Official Websites of Supreme Election Council, access May 2019
Political Competition	The rate of vote differences between first and second candidates in the last mayoral election	Official Websites of Supreme Election Council, access May 2019

Source: Own Elaboration

FINDINGS

General Overview

There are 81 provinces, and each province has one municipality in Turkey. Thirty of the municipalities in these provinces are metropolitan municipalities, and 51 are provincial municipalities. Due to cancellation of the mayoral elections in Istanbul, the metropolitan municipality of Istanbul was not included in the study. It was found that 29 mayors of metropolitan municipalities and 48 mayors of provincial municipalities have a Twitter account. However, it was noticed that all mayors who have a Twitter account do not use Twitter actively. In this study, Twitter accounts of mayors who tweeted less than ten tweets during the 30 days (17 April-16 May) were considered passive accounts. In this context, it was seen that the Twitter accounts of 9 mayors were passive. As a result, 29 mayors of the metropolitan municipality and 39 mayors of the provincial municipality were included in the study.

Engagement scores of all mayors are illustrated in Table 3. The average of the engagement level of mayors is 23.29, and 65% of mayors are below the average. The mayor with the highest number of followers is Mansur Yavaş (Ankara) with

Table 3. Engagement scores of mayors

Name of Municipality	Name of Mayor	Political Parties	Number of Followers	Number of Tweets	Engagement
Municipality of Sinop	Barış Ayhan	CHP	1518	35	113,89
Metropolitan Municipality of Mardin	Ahmet Türk	HDP	22829	11	81,14
Municipality of Adıyaman	Süleyman Kilinç	AK Party	3447	10	80,62
Municipality of Karaman	Savaş Kalaycı	MHP	1003	30	77,37
Municipality of Kırşehir	Selahattin Ekicioğlu	CHP	2905	22	75,43
Municipality of Kilis	Mehmet Abdi Bulut	AK Party	530	57	72,39
Municipality of Bingöl	Erdal Arikan	AK Party	4000	37	60,07
Municipality of Yozgat	Celal Köse	AK Party	1632	19	55,73
Municipality of Afyonkarahisar	Mehmet Zeybek	AK Party	2092	28	51,85
Municipality of Amasya	Mehmet Sarı	MHP	2777	44	50,05
Municipality of Giresun	Aytekin Şenlikoğlu	AK Party	2804	59	41,97
Metropolitan Municipality of Diyarbakır	Adnan Selçuk Mızraklı	HDP	45689	31	40,51
Metropolitan Municipality of İzmir	Tunç Soyer	CHP	381403	66	36,54
Metropolitan Municipality of Kocaeli	Tahir Büyükakın	AK Party	18195	66	33,41
Municipality of Bitlis	Nesrullah Tanğlay	AK Party	3919	63	32,91
Municipality of Iğdır	Yaşar Akkuş-Eylem Çelik	HDP	1868	22	32,56
Municipality of Erzincan	Bekir Aksun	MHP	3364	14	32
Municipality of Batman	Mehmet Demir	HDP	2342	59	28,37
Municipality of Nevşehir	Rasim Arı	AK Party	26020	226	27,44
Municipality of Isparta	Şükrü Başdeğirmen	AK Party	1908	50	26,86
Metropolitan Municipality of Adana	Zeydan Karalar	CHP	78403	105	26,25
Municipality of Elazığ	Şahin Şerifoğulları	AK Party	7083	148	25,62
Municipality of Kastamonu	R. Galip Vidinlioğlu	MHP	1931	72	24,24
Municipality of Zonguldak	Ömer Selim Alan	AK Party	1712	103	23,63
Metropolitan Municipality of Kahramanmaraş	Hayrettin Güngör	AK Party	7635	139	22,19
Municipality of Bolu	Tanju Özcan	CHP	99227	29	21,78
Municipality of Tunceli	Fatih Mehmet Maçoğlu	TKP	350770	14	21,28
Municipality of Niğde	Emrah Özdemir	AK Party	2600	149	20,83
Municipality of Şırnak	Mehmet Yarka	AK Party	3815	34	19,8
Metropolitan Municipality of Eskişehir	Yılmaz Büyükerşen	CHP	281781	64	19,32
Metropolitan Municipality of Samsun	Mustafa Demir	AK Party	2847	176	19,26
Metropolitan Municipality of Sakarya	Ekrem Yüce	AK Party	9761	130	19,07
Metropolitan Municipality of Muğla	Osman Gürün	CHP	10883	54	15,59
Metropolitan Municipality of Ankara	Mansur Yavaş	CHP	2060461	49	15,22
Municipality of Çorum	Halil İbrahim AŞGIN	AK Party	1353	437	14,56
Municipality of Kırklareli	Mehmet Siyam Kesimoğlu	CHP	61027	132	13,89
Metropolitan Municipality of Aydın	Özlem Çerçioğlu	CHP	96794	56	13,72
Metropolitan Municipality of Trabzon	Murat Zorluoğlu	AK Party	31816	55	12,43
Metropolitan Municipality of Mersin	Vahap Seçer	CHP	23851	181	12,37
Metropolitan Municipality of Ordu	Mehmet Hilmi Güler	AK Party	10803	155	12,3
Metropolitan Municipality of Malatya	Selahattin Gürkan	AK Party	12092	202	12,24
Metropolitan Municipality of Van	Bedia Özgökçe Ertan	HDP	32398	34	10,71
Metropolitan Municipality of Şanlıurfa	Zeynel Abidin Beyazgül	AK Party	34977	241	10,17
Metropolitan Municipality of Konya	Uğur İbrahim Altay	AK Party	61129	84	9,88
Municipality of Çanakkale	Ülgür Gökhan	CHP	14926	112	9,45
Municipality of Ağrı	Savcı Sayan	AK Party	962666	65	8,6
Metropolitan Municipality of Hatay	Lütfü Savaş	CHP	44066	119	8,22
Municipality of Karabük	Rafet Vergili	MHP	2536	157	7,8
Municipality of Kars	Ayhan Bilgen	HDP	314300	73	7,28
Metropolitan Municipality of Antalya	Muhittin Böcek	CHP	116976	107	7,08
Municipality of Muş	Feyat Asya	AK Party	7429	92	6,95

continued on following page

Table 3. Continued

Name of Municipality	Name of Mayor	Political Parties	Number of Followers	Number of Tweets	Engagement
Metropolitan Municipality of Balıkesir	Yücel Yılmaz	AK Party	22327	95	5,94
Metropolitan Municipality of Tekirdağ	Kadir Albayrak	CHP	11241	184	5,55
Metropolitan Municipality of Kayseri	Memduh Büyükkılıç	AK Party	15839	350	5,3
Municipality of Osmaniye	Kadir Kara	MHP	3765	73	4,99
Municipality of Burdur	Ali Orkun Ercengiz	CHP	11880	120	4,85
Municipality of Tokat	Eyüp Eroğlu	AK Party	15145	148	4,48
Municipality of Gümüşhane	Ercan Çimen	AK Party	8100	69	4,47
Metropolitan Municipality of Manisa	Cengiz Ergün	MHP	14146	127	3,91
Metropolitan Municipality of Bursa	Alinur Aktaş	AK Party	75919	181	3,77
Municipality of Sivas	Hilmi Bilgin	AK Party	26359	70	3,61
Metropolitan Municipality of Denizli	Osman Zolan	AK Party	44089	125	3,29
Municipality of Bartın	Cemal Akın	MHP	3861	22	3,28
Municipality of Edirne	Recep Gürkan	CHP	73968	195	2,75
Metropolitan Municipality of Erzurum	Mehmet Sekmen	AK Party	71314	168	1,68
Municipality of Yalova	Vefa Salman	CHP	36734	150	1,55
Municipality of Düzce	Faruk Özlü	AK Party	89818	34	1,03
Metropolitan Municipality of Gaziantep	Fatma Şahin	AK Party	1376762	131	0,39

Source: Own Elaboration

followers 2060461, and the mayor with the smallest number of followers is Mehmet Abdi Bulut (Kilis) with followers 530. In addition to these, the mayor who is the most active Twitter user is Halil İbrahim Aşgın (Çorum) with 437 tweets.

The Relationship between Factors and Engagement Level

Various statistical analysis techniques were used to determine the factors associated with the engagement level of the mayors. The techniques and results are illustrated in Table 4. According to the analysis, there is no relationship between the personal traits of mayors (gender (RQ1), age (RQ2), and education level (RQ3)) and the

Table 4. Relationship between factors and engagement

Factors	Engagement	Method
Gender	(U=54.0, p= .194)	Mann-Whitney U
Age	[r= -.072, n=68, p=.557]	Correlation Analysis
Education Background	[F (3, 63)=.802, p=.4.97]	ANOVA
Term	[F (3, 64)=2.05 p=.115]	ANOVA
Status of Municipality	[t (66) = -2.20, p= .031]	t test
Region of Municipality	[F (6, 61)=1.74 p=.127]	ANOVA
Political Parties of Mayor	[F (3, 63)=.437, p=.727]	ANOVA
Political Competition	[r= -.123, n=68, p=.316]	Correlation Analysis

Source: Own Elaboration

engagement level. Similarly, there is no relationship between political context (competition (RQ6), party (RQ7), and terms (RQ8)) and engagement level of mayors.

Nevertheless, as a result of the tests to analyse the relationship between the level of engagement and the characteristics of the municipality, it was found that there is no significant relationship between the geographical region (RQ5) and the engagement level of the municipality, but there is a significant relationship between the status of municipality (RQ6) and engagement level of mayors. The mayors of provincial municipalities have more engagement score than the mayors of metropolitan municipalities.

CONCLUSION

According to the findings, it is shown that most of the mayors (96%) in Turkey have a Twitter account. However, some of them (11%) do not use Twitter actively. Furthermore, the active Twitter user differ widely in terms of their activity (10 tweets to 437 tweets). Sobaci and Karkin (2013) found that 43 out of 81 mayors had a Twitter account in Turkey. Therefore, it can be said that the rate of mayors' awareness of Twitter in 2019 increased by 81% compared to 2013. Also, the rate of increase much more than a rate of increase of active Twitter users worldwide from 2013 to 2019. Because, the active Twitter users worldwide increased by 37% (from 241 to 330 million) (Statista, 2019). One of the most significant reasons for the rate of increase can be that mayors comprehended the opportunities offered by social media for political communication.

Although Szmigiel-Rawska et al. (2018) found that younger mayor is more likely to be successful, the result of our analysis reveals that there is no relationship personal traits of mayors and engagement. Today, the power of social media is noticed by all political actors. So, social media accounts of the majority of politicians are managed by professionals. The lack of relationship between personal traits of mayors and engagement level can be explained by these factors. According to results, as well as personal traits, political parties of mayors and terms of mayors don't influence the engagement level. These results can be explained like personal traits by the proliferation of professional social media management.

The other factor that was found to be unrelated to the engagement level is geographical regions of municipalities. It was expected that the engagement level of mayors varies in different geographical regions due to the differences in levels of economic development. But findings didn't support this hypothesis. It is possible to argue that due to the proliferation of low-priced internet devices, especially smartphones, there is no significant relationship between economic development and engagement level. The last factor which has no influence on the engagement level is political competition. Although social media are effective tools for the permanent political campaign, the effects of political competition emerge clearly in the election period. However, this research focused on non-election period. Therefore, the lack of significant relationship between political competition and engagement level can be explained by this situation.

Previous studies found that there is no relationship between municipality size and engagement level (Bonson et al. 2017; 2019). However, our findings show that the status of municipalities influences the engagement level of mayors. As mentioned before, the status of municipalities is directly related to the population of municipalities in Turkey. Therefore, the mayors of small municipalities have more engagement than mayors of big municipalities. It can be explained by the argument that since followers see their opinion matters, which encourages them to interact more with the mayors, they feel closer to the mayors in smaller municipalities (Bonson et al., 2019).

Despite the significant findings, this study has several limitations. Firstly, it focused on only one social media platform for measuring the engagement level. Secondly, this study includes only the mayors of provincial and metropolitan municipalities. In addition to these, there are small-town municipalities and district municipalities in Turkey. Lastly, this study focus on a single country and period of time. Therefore, in the literature of social media in politics, there is needed the more empirical studies which focus on determinants of social media success of political actors in national and local levels. In this context, future studies can analyse the determinants of the engagement level of mayors of district municipalities. Moreover, the engagement level of mayors in different countries can be analysed comparatively.

REFERENCES

Agostino, D., & Arnaboldi, M. (2016). A measurement framework for assessing the contribution of social media to public engagement: An empirical analysis on Facebook. *Public Management Review*, *18*(9), 1289–1307. doi:10.1080/1471903 7.2015.1100320

Bebić, D., & Vučković, M. (2013). Facebook usage by mayors in Central and Southeastern Europe. *Medijske studije. Mediaeval Studies*, *4*(8), 32–44.

Bode, L. (2012). Facebooking it to the polls: A study in online social networking and political behavior. *Journal of Information Technology & Politics*, *9*(4), 352–369. doi:10.1080/19331681.2012.709045

Bonsón, E., Perea, D., & Bednárová, M. (2019). Twitter as a tool for citizen engagement: An empirical study of the Andalusian municipalities. *Government Information Quarterly*, *36*(3), 480–489. doi:10.1016/j.giq.2019.03.001

Bonsón, E., & Ratkai, M. (2013). A set of metrics to assess stakeholder engagement and social legitimacy on a corporate Facebook page. *Online Information Review*, *37*(5), 787–803. doi:10.1108/OIR-03-2012-0054

Bonsón, E., Ratkai, M., & Royo, S. (2016). Facebook use in Western European local governments: An overall view. In M. Z. Sobaci (Ed.), *Social media and local governments* (pp. 59–77). Cham: Springer. doi:10.1007/978-3-319-17722-9_4

Bonsón, E., Royo, S., & Ratkai, M. (2015). Citizens' engagement on local governments' Facebook sites. An empirical analysis: The impact of different media and content types in Western Europe. *Government Information Quarterly*, *32*(1), 52–62. doi:10.1016/j.giq.2014.11.001

Bonsón, E., Royo, S., & Ratkai, M. (2017). Facebook practices in Western European municipalities: An empirical analysis of activity and citizens' engagement. *Administration & Society*, *49*(3), 320–347. doi:10.1177/0095399714544945

Carlson, T., & Strandberg, K. (2008). Riding the Web 2.0 wave: Candidates on YouTube in the 2007 Finnish national elections. *Journal of Information Technology & Politics*, *5*(2), 159–174. doi:10.1080/19331680802291475

Chu, H., & Xu, C. (2009). Web 2.0 and its dimensions in the scholarly world. *Scientometrics*, *80*(3), 717–729. doi:10.100711192-008-2103-y

Criado, J. I., Martínez-Fuentes, G., & Silván, A. (2012). Social media for political campaigning. The use of Twitter by Spanish Mayors in 2011 local elections. In C. G. Reddick & S. K. Aikins (Eds.), *Web 2.0 technologies and democratic governance: Political, Policy and Management Implications* (pp. 219–232). New York, NY: Springer. doi:10.1007/978-1-4614-1448-3_14

Criado, J. I., Sandoval-Almazan, R., & Gil-Garcia, J. R. (2013). Government innovation through social media. *Government Information Quarterly*, *30*(4), 319–326. doi:10.1016/j.giq.2013.10.003

De Zúñiga, H. G., Copeland, L., & Bimber, B. (2014). Political consumerism: Civic engagement and the social media connection. *New Media & Society*, *16*(3), 488–506. doi:10.1177/1461444813487960

Eagly, A. H., & Johnson, B. T. (1990). Gender and leadership style: A meta-analysis. *Psychological Bulletin*, *108*(2), 233–256. doi:10.1037/0033-2909.108.2.233

Eltantawy, N., & Wiest, J. B. (2011). The Arab spring| Social media in the Egyptian revolution: Reconsidering resource mobilization theory. *International Journal of Communication*, *5*, 1207–1224.

Gandía, J. L., & Archidona, M. C. (2008). Determinants of web site information by Spanish city councils. *Online Information Review*, *32*(1), 35–57. doi:10.1108/14684520810865976

García, A. C., & García-García, J. (2010). Determinants of online reporting of accounting information by Spanish local government authorities. *Local Government Studies*, *36*(5), 679–695. doi:10.1080/03003930.2010.506980

Hansen, K. M., & Kosiara-Pedersen, K. (2014). Cyber-campaigning in Denmark: Application and effects of candidate campaigning. *Journal of Information Technology & Politics*, *11*(2), 206–219. doi:10.1080/19331681.2014.895476

Haro-de-Rosario, A., Sáez-Martín, A., & del Carmen Caba-Pérez, M. (2018). Using social media to enhance citizen engagement with local government: Twitter or Facebook? *New Media & Society*, *20*(1), 29–49. doi:10.1177/1461444816645652

İkiz Ozdeşim, O., Sobaci, M. Z., Yavuz, N., & Karkin, N. (2014, October). Political use of Twitter: The case of metropolitan mayor candidates in 2014 local elections in Turkey. In *Proceedings of the 8th international conference on theory and practice of electronic governance* (pp. 41-50). ACM.

Kalnes, Ø. (2009). Norwegian parties and Web 2.0. *Journal of Information Technology & Politics*, *6*(3-4), 251–266. doi:10.1080/19331680903041845

Kearney, M. W. (2016). *rtweet: Collecting Twitter data. Comprehensive R archive network*. Retrieved from https://cran.r-project.org/package=rtweet

Larsson, A. O. (2018). Small is the new big–at least on Twitter: A diachronic study of Twitter use during two regional Norwegian elections. In M. Z. Sobaci & İ. Hatipoğlu (Eds.), *Sub-national democracy and politics through social media* (pp. 169–182). Cham: Springer. doi:10.1007/978-3-319-73386-9_9

Larsson, A. O., & Moe, H. (2014). Triumph of the underdogs? Comparing Twitter use by political actors during two Norwegian election campaigns. *SAGE Open*, *4*(4), 1–13. doi:10.1177/2158244014559015

Lev-On, A. (2018). Perceptions, uses, visual aspects, and consequences of social media campaigning: Lessons from municipal Facebook campaigning, Israel 2013. In M. Z. Sobaci & İ. Hatipoğlu (Eds.), *Sub-national democracy and politics through social media* (pp. 149–168). Cham: Springer. doi:10.1007/978-3-319-73386-9_8

Morris, M. G., & Venkatesh, V. (2000). Age differences in technology adoption decisions: Implications for a changing work force. *Personnel Psychology*, *53*(2), 375–403. doi:10.1111/j.1744-6570.2000.tb00206.x

O'Reilly, T. (2007). What is web 2.0: Design patterns and business models for the next generation of software. *Communications & Stratégies*, *65*(1), 17–37.

Paek, H. J., Hove, T., Jung, Y., & Cole, R. T. (2013). Engagement across three social media platforms: An exploratory study of a cause-related PR campaign. *Public Relations Review*, *39*(5), 526–533. doi:10.1016/j.pubrev.2013.09.013

Rowe, G., & Frewer, L. J. (2005). A typology of public engagement mechanisms. *Science, Technology & Human Values, 30*(2), 251–290. doi:10.1177/0162243904271724

Sheedy, A., MacKinnon, M. P., Pitre, S., & Watling, J. (2008). *Handbook on citizen engagement: Beyond consultation*. Ontario: Canadian Policy Research Networks.

Skoric, M. M., Zhu, Q., Goh, D., & Pang, N. (2016). Social media and citizen engagement: A meta-analytic review. *New Media & Society*, *18*(9), 1817–1839. doi:10.1177/1461444815616221

Sobaci, M. Z., & Hatipoğlu, İ. (2017, May). *Measuring the engagement level of political parties with public on Facebook: The case of Turkey. In 2017 Conference for E-Democracy and Open Government (CeDEM)* (pp. 209–216). IEEE.

Sobacı, M. Z., Hatipoğlu, İ., & Korkmaz, M. F. (2018). *The effect of post type and post category on citizen interaction level on Facebook: The case of metropolitan and provincial municipalities in the Marmara Region of Turkey. In Sub-national democracy and politics through social media* (pp. 91–105). Cham: Springer.

Sobaci, M. Z., & Karkin, N. (2013). The use of twitter by mayors in Turkey: Tweets for better public services? *Government Information Quarterly*, *30*(4), 417–425. doi:10.1016/j.giq.2013.05.014

Statista. (2019). *Number of monthly active Twitter users worldwide from 1st quarter 2010 to 1st quarter 2019 (in millions)*. Retrieved July 5, from https://www.statista.com/statistics/282087/number-of-monthly-active-twitter-users/

Strandberg, K. (2013). A social media revolution or just a case of history repeating itself? The use of social media in the 2011 Finnish parliamentary elections. *New Media & Society*, *15*(8), 1329–1347. doi:10.1177/1461444812470612

Szmigiel-Rawska, K., Łukomska, J., & Tavares, A. F. (2018, April). Social Media Activity and Local Civic Engagement in Poland. In *Proceedings of the 11th International Conference on Theory and Practice of Electronic Governance* (pp. 279-287). ACM. 10.1145/3209415.3209516

Tavares, A. F., & da Cruz, N. F. (2017). Explaining the transparency of local government websites through a political market framework. *Government Information Quarterly*, 101249. doi:10.1016/j.giq.2017.08.005

Triantafillidou, A., Lappas, G., Kleftodimos, A., & Yannas, P. (2018). Attack, Interact, and Mobilize: Twitter Communication Strategies of Greek Mayors and their Effects on Users' Engagement. In M. Z. Sobaci & İ. Hatipoğlu (Eds.), *Sub-national democracy and politics through social media* (pp. 65–89). Cham: Springer. doi:10.1007/978-3-319-73386-9_4

Triantafillidou, A., Lappas, G., & Yannas, P. (2015). Facebook engagement and Greek local municipal governments. In *Proceedings of the International Conference for E-Democracy and Open Governement* (pp. 39-52). Krems: Donau-Universitat.

Valenzuela, S., Park, N., & Kee, K. F. (2009). Is there social capital in a social network site?: Facebook use and college students' life satisfaction, trust, and participation. *Journal of Computer-Mediated Communication*, *14*(4), 875–901. doi:10.1111/j.1083-6101.2009.01474.x

Vergeer, M., Hermans, L., & Sams, S. (2013). Online social networks and micro-blogging in political campaigning: The exploration of a new campaign tool and a new campaign style. *Party Politics*, *19*(3), 477–501. doi:10.1177/1354068811407580

Warren, A. M., Sulaiman, A., & Jaafar, N. I. (2014). Social media effects on fostering online civic engagement and building citizen trust and trust in institutions. *Government Information Quarterly*, *31*(2), 291–301. doi:10.1016/j.giq.2013.11.007

Welp, Y., Capra, P., & Freidenberg, F. (2018). Politics and digital media: An exploratory study of the 2014 subnational elections in Ecuador. In M. Z. Sobaci & İ. Hatipoğlu (Eds.), *Sub-national democracy and politics through social media* (pp. 207–222). Cham: Springer. doi:10.1007/978-3-319-73386-9_11

Williams, C. B., & Gulati, G. J. J. (2013). Social networks in political campaigns: Facebook and the congressional elections of 2006 and 2008. *New Media & Society*, *15*(1), 52–71. doi:10.1177/1461444812457332

Zhang, W., Johnson, T. J., Seltzer, T., & Bichard, S. L. (2010). The revolution will be networked: The influence of social networking sites on political attitudes and behavior. *Social Science Computer Review*, *28*(1), 75–92. doi:10.1177/0894439309335162

Chapter 8
Explaining Government Crowdsourcing Decisions:
A Theoretical Model

Nilay Yavuz
https://orcid.org/0000-0002-1673-6309
Middle East Technical University, Turkey

Naci Karkın
https://orcid.org/0000-0002-0321-1212
Pamukkale University, Turkey

Ecem Buse Sevinç Çubuk
https://orcid.org/0000-0002-1679-1746
Aydin Adnan Menderes University, Turkey & Delft University of Technology, The Netherlands

ABSTRACT

Crowdsourcing online has been popularly utilized especially among business organizations to achieve efficiency and effectiveness goals and to obtain a competitive advantage in the market. With the governments' increasing interest in using information and communication technologies for a variety of purposes, including generation of public value(s) and innovative practices, online crowdsourcing has also entered into the public administration domain. Accordingly, studies have investigated critical success factors for governmental crowdsourcing, or explored citizen participation in crowdsourcing activities in case studies. However, governmental decision to adopt online crowdsourcing as innovation has not been sufficiently examined in the extant literature. The objective of this chapter is to propose a theoretical model that explains the government adoption of crowdsourcing. Based on the review of case studies on governmental crowdsourcing, an integrated theoretical model of factors affecting government crowdsourcing decisions is developed.

DOI: 10.4018/978-1-7998-1526-6.ch008

INTRODUCTION

With the ever-growing advancements in ICTs, there emerge many opportunities for governments to innovate public service delivery, to improve democratic outcomes, and to undertake administrative reforms. One of the digitally enabled government innovations that have been gaining more attention recently is government crowdsourcing. Crowdsourcing can be defined as "the act of an organization taking a function once performed by an organization's own employees and outsourcing it to people outside the organization (crowd) through an open call online" (Liu, 2017, p. 656; Howe, 2006a). In public context, main examples of crowdsourcing include involving citizens in the production of public services, inviting the public to solve public problems, and incorporating public participation into policy making (Nam, 2012; Liu, 2017). It is argued that crowdsourcing has the potential to provide solutions to persistent or emergent issues and problems that may not be met by traditional bureaucratic efforts (Bommert, 2010). Crowdsourcing platforms "can empower citizens, create legitimacy for the government with the people, and enhance the effectiveness of public services and goods" (Liu, 2017, p. 656). In addition, crowdsourcing can be very functional to employ the public at large for searching for new ways and methods to conduct the business, particularly at lower costs when compared to paying traditional employees (Howe, 2006b). Since crowdsourcing can be operationalized using online networks and communities through the intra- and internet, it eliminates time and space as the limiting elements together with the associated costs that public institutions are assumed to bear when aiming to address societal issues.

Nonetheless, the initial experience of governments with these online platforms has been a trial-and-error process (Brabham, 2009). While the potential benefits and challenges mentioned earlier -to some extent- explain governments' decisions to crowdsource online, there is a need to adopt a more holistic approach in understanding governmental adoption of crowdsourcing, and consider a variety of factors in explaining this process, including the decision to use information and communication technologies.

In a study that conducts a systematic review of crowdsourcing decisions in the business sector, Thuan, Antunes and Johnstone (2016, p.48) assert that "there is lack of a commonly accepted list of factors that affect the decision to crowdsource" and "the overall picture on the crowdsourcing decision is still unveiled". Existing research on government crowdsourcing has largely focused on dispersed case studies or the citizens' perspective. However, factors that affect governments to engage in online crowdsourcing are not comprehensively explored in the literature.

In the light of this gap, the objective of this study is to identify factors related to government online crowdsourcing decisions. By conducting a review of the case studies on government online crowdsourcing, the chapter proposes an integrated theoretical model of factors affecting government crowdsourcing decisions, including individual perceptions, organizational factors, and environmental factors. According to this framework, some policy recommendations for government crowdsourcing decisions are made.

The paper is organized as follows. The first section discusses crowdsourcing concept and presents a theoretical framework for the study. In the next section, methodology for the study is explained, and findings from the review of the case studies on governmental online crowdsourcing are presented. Based on this review, a theoretical model is developed for government online crowdsourcing adoption. Finally, conclusions discuss research and policy implications of the model.

CROWDSOURCING AND ITS APPLICATION IN THE PUBLIC SECTOR

First coined by Jeff Howe and Mark Robinson in 2006, crowdsourcing can be defined as "deliberate blend of bottom-up, open, creative process with top-down organizational goals" (Brabham, 2013a, xv-xvii). Intrinsically including benefits and difficulties, Howe (2008, p. 18) argues that crowdsourcing "is just a rubric for a wide range of activities. Its adaptability is what makes it pervasive and powerful. But this very flexibility makes the task of defining and categorizing crowdsourcing a challenge". In its earlier form, Howe (2006a, p. 1) had defined the term as "representing the act of a company or institution taking a function once performed by employees and outsourcing it to an undefined (and generally large) network of people in the form of an open call". It should be noted that the term crowdsourcing has many definitions, even conflicting in the same paper, in the studies of the relevant literature (Brabham, 2013). It is also used interchangeably with coproduction. For this very reason, Brabham (2013, p. 7) points to a systematic literature analysis by Estellés-Arolas and González-Ladrón-de-Guevara who come up with a definition of crowdsourcing as "a type of participative online activity in which an individual, an institution, a non-profit organization, or company proposes to a group of individuals of varying knowledge, heterogeneity, and number, via a flexible open call, the voluntary undertaking of a task" (2012, p. 197). Crowd, on the other hand, is usually conceptualized as "anybody who has access to the Internet and is aware of the task" (Aitamurto and Landemore, 2016: 177).

Crowdsourcing has many benefits; particularly, cost and time-effective collection of novel information from a large community, and transforming the interactions between public and the government that might result in innovating government business and producing public values (Brabham, 2009). In addition, it encloses potential shareholders and their possible material and immaterial assets directed to solve a particular, or a persistent issue (Afzalan, Evans-Cowley, and Mirzazad, 2015). For this very reason, it embodies production of public values and innovation per se. Regarding the ICTs use, crowdsourcing can be operationalized via online networks and communities through the intra- and Internet. Noting that online communities and online crowdsourcing are not identical to each other (Borst, 2010), using online crowdsourcing as an ICT tool might help outsourcing ideas and designs through an innovative way (Seltzer and Mahmoudi, 2013).

According to Liu (2017, p. 657), the relevant literature clearly shows that crowdsourcing might facilitate novel relationships, or transform existing interactions between government and citizens. Nonetheless there are also some challenges in the implementation process. (Garcia, Vivacqua, & Tavares, 2011; Prpić, Taeihagh, and Melton, 2015; Hansson, Muller, Aitamurto, Light, Mazarakis, Gupta, & Ludwig, 2016; Fitzgerald, McCarthy, Carton, O Connor, Lynch, & Adam, 2016; Sivarajah, Weerakkody, Waller, Lee, Irani, Choi, Morgan, & Glikman, 2016). Such challenges particularly include bureaucratic resistance to share the competences with various stakeholders including ordinary citizens (Minner, 2015; Slotterback, 2011), inadequate governmental capacity (Klosterman, 2013; Townsend, 2013) and expertise with regard to the acceptance and use of ICTs (Brabham, 2009; Saad-Sulonen, 2012), and lack of synergy between citizens and government (Aladalah, Cheung and Lee, 2016).

Research gives some successful examples of crowdsourcing. Nemec, Svidroňová, Meričková, Klimovský (2017, pp. 287-288) exemplifies the case of "Conciliation councils" that were created as citizen initiatives, where association of citizens assumed mediating roles for solving all kinds of conflicts, in some multiethnic regions as towns of Levice, Nove Zamky, Kežmarok, Rimavska Sobota, Prešov and their surroundings. In these councils "the citizens act as initiators, (co)designers, and (co)implementers" (Bertot, Estevez, & Janowski 2016, p. 218). Bommert (2010) counts some positive examples of public sector innovation cases as Open University and the National Literacy Strategy in the UK, and Ford Foundation's Innovations in American Government program in the US.

Similarly, Bertot et al. (2016) gives the example of the Municipality of Amsterdam for its provision of a crowdsourcing platform enabling co-creation of applications and the delivery of public value with citizens. According to them, "through the platform, citizens are encouraged to publish ideas about new practical applications that can add value to Amsterdam dwellers, and other citizens can comment on or discuss the ideas" including the proponents (2016, p. 218).

THEORETICAL FRAMEWORK OF ONLINE CROWDSOURCING

As noted above, the aim of this paper is to develop a theoretical model of factors affecting governmental decision to crowdsource online. For this very purpose, first, existing theories related to crowdsourcing are to be reviewed. As crowdsourcing activity itself is commonly regarded as a novel approach (Howe, 2006b), especially when conducted along with the use of Internet technologies such as social media or other websites, it is possible to think of crowdsourcing adoption in terms of the adoption of an innovation that involves ICT use. The term "innovation" refers to "the successful introduction of a new thing or method . . .the embodiment, combination, or synthesis of knowledge in original, relevant, valued new products, processes, or services" (Luecke & Katz, 2003, p. 2). In line with this, one of the theoretical frameworks that will be utilized in this study is the adoption of innovation framework, which is reviewed below. The framework also highlights the factors that explain the decision to use of ICTs, as it is relevant to the online government crowdsourcing process.

Innovation / Technology Adoption Theories

There are many theories of innovation/technology adoption. Technology Acceptance Model (TAM) proposed by Davis, Bagozzi and Warshaw (1989) is a well-known theory that explains users' intention and behavior of IT by perceived usefulness and perceived ease of use of the technology. Perceived usefulness refers to the extent to which a person believes that a system will enhance his/her performance, and perceived ease of use is the extent to which a person believes that adopting the innovation will be free of effort and risk. Davis et al. (1989) proposed that user attitudes, that is, feelings of favorableness or unfavorableness towards performing a behavior, are determined by perceived usefulness and perceived ease of use, which in turn affects behavioral intention to use IT and the actual usage behavior.

According to Theory of Reasoned Action, in addition to perceived usefulness and ease of use, "subjective norm" also affects innovation adoption decisions. Fishbein and Ajzen (1975, p. 302) defines subjective norm as "a person's perception that most people who are important to him think he should or should not perform the behavior in question". In other words, this theory takes into consideration social influences as a factor that affects innovation adoption decisions.

Research shows that besides these perceptions, internal and external constraints may affect individuals' adoption of innovations. For example, Theory of Planned Behavior (TPB) hypothesizes that adoption behavior is a direct function of behavioral intention and "perceived behavioral control" which refers to the perceptions of internal and external constraints on behavior. Ajzen (1991) points out that perceived behavioral control is about the beliefs of individuals about access to resources and opportunities needed to perform a behavior, or alternatively, to the internal and external constraints that may impede realization of the behavior. Taylor and Todd (1995) argue that there are two dimensions of this concept. One is "facilitating conditions" (Triandis, 1979) such as time, money, or other resources needed to perform a behavior, and the other is self-efficacy, which is an individual's self-confidence in his/her ability to perform a behavior (Bandura, 1977).

Similarly, presence of knowledge and skills and availability of resources may also affect implementation of innovation (Ely, 1993). It is essential that the people who will use the innovation possess the technical knowledge and skills. In addition, Farr and Ford (1990) assert that, individuals' self-efficacy beliefs are as important as the actual presence of skills. They argue that self-efficacy would influence a person's decision to put forth the effort to adopt an innovation. In addition to presence of skills, appropriate and adequate resources including finances, hardware, and software must be available and accessible to users to successfully implement the innovation.

Rogers (1995) argues that certain characteristic of innovations help explain their acceptance among individuals. These are simplicity, trialability, observability, and relative advantage characteristics of innovations

Simplicity refers to whether the innovation is easy to understand, maintain, and use. The more users think that the innovation is inherently complex, difficulty to understand and integrate into daily practices, the slower its adoption will be (Rogers 1995; DiMaggio and Cohen 2005).

Another characteristic of the innovation, trialability refers to the opportunity for individuals to try out the innovation on a limited basis and change their adoption decision if desired. When an innovation is trialable, individuals may be less likely to perceive the adoption decision as risky and more likely to adopt. Observability feature of an innovation is described as the results of the innovation being visible to the individuals for them to observe how it works and evaluate the consequences (Rogers, 1995). It may affect the perceived usefulness and ease of use of the innovation.

Regarding the relative advantage of innovation, a comparison is made with the available alternatives to see whether the innovation is more economical, socially prestigious, more convenient and more satisfying, and adoption occurs in case of an extra benefit (Rogers, 1995).

To summarize, theories of innovation adoption indicate that perceived usefulness, perceived ease of use, constraints, social influences, and relative advantage of the technology are the main constructs that relate to the decision to adopt innovations.

Organizational and Environmental Factors Related to Online Crowdsourcing

Tornatzky and Fleischer (1990) in their technology–organization–environment (TOE) framework proposed that besides the technological context, the organizational context and the external environmental context affect innovation adoption decisions. Some of the technology adoption theories reviewed above also highlight the importance of organizational and external influences. Accordingly, this section reviews research on the effects of organizational and environmental factors on IT innovations, and crowdsourcing in particular.

Organizational Context

Prior studies showed that the nature of the task is an important factor in crowdsourcing decisions (Thuan et al., 2016). According to Zhao and Zhu (2014), the "crowd" is appropriate for certain tasks, but not for others. For example, tasks that involve privacy and security issues, and intellectual property features may not be suitable for crowdsourcing (Muntés-Mulero, Paladini, Manzoor, Gritti, Larriba-Pey, & Mijnhardt, 2013). In addition, because crowdsourcing activities usually involve the use of the Internet technology, the tasks in question should be the ones that can be managed through the Internet (Doan, Ramakrishnan, & Halevy, 2011). According to research, other required characteristics of the tasks are that, they should not need complex training for the crowd and be performed without significant need for interaction (Muntés-Mulero et al., 2013); they should be well-defined, have a clear-scope, and can be broken down into reasonable tasks to be performed by the public (Lloret, Plaza, & Aker, 2012; Seltzer & Mahmoudi, 2013; Zogaj, Bretschneider, Leimeister, 2014; Afuah & Tucci, 2012 ; Prpić, Taeihagh, & Melton 2015). Crowdsourcing decisions can also be based on whether the task has any urgency and whether the organization can significantly save time and cost by getting input from the crowd for that urgency instead of doing the task by itself.

As another organizational factor, innovation is very conditional to commitment, leadership and support at the managerial level (Gil-Garcia et al., 2016, p. 527). It is argued that top management provides vision, guideline and support to build the environment necessary for IT innovations (Lee & Kim, 2007).

For crowdsourcing adoption consideration as an innovation, other organizational factors include sufficient budget to cover crowdsourcing activities, availability of a crowdsourcing platform (infrastructure), the availability of expertise to manage the crowdsourcing activity and risks and self-efficacy of the manager, and an organizational culture that promotes employees' commitment (Thuan et al., 2016: 58).

Environmental Context

Previous research shows that innovation decisions may be related to environmental factors including political influences, practices of other public organizations, and the citizens' interest (Wang & Lo, 2016). Based on the "concept of coercive pressures from institutional theory", Wang and Lo (2016, p. 84) point out that external pressures coming from central and local governments may influence government agency decisions to adopt an innovation, including open government data. Similarly, they argue that innovation decisions may also be affected by the best practices employed by other government agencies. In other words, decisions to innovate may be motivated by what other organizations are doing and whether or not they are successful.

For crowdsourcing practices, one of the defining elements is the availability of the "crowd" that is the public to undertake the task (Thuan et al., 2016). Research indicates that availability of the public to procure for a task using the Internet positively affects the decision to crowdsource (Malone, Laubacher, & Dellarocas, 2010; Thuan et al., 2016). However, in engaging with government to contribute to the production of public value, citizens may lack the opportunities, abilities and motivation. Not every layer of society, or every region could be equally interested in digital involvement due to various reasons ranging from economics to aspiration, to citizen distrust in government (Meijer, 2015). Therefore, the number and other characteristics of the participants may affect crowdsourcing decisions.

METHODOLOGY OF THE RESEARCH

In order to identify the factors related to government online crowdsourcing decisions and develop an integrated theoretical model, this study used the systematic literature review methodology suggested by Okoli and Schabram (2010). The review included a selection of peer-reviewed articles including government crowdsourcing case studies published in academic journals during the last decade (2010 to 2019). The purpose of selecting this time frame was to capture the years that the use of ICTs in government was becoming widespread. The initial article search was conducted by selecting the following Boolean expressions in the "topic" section of Web of Science Core Collection (WoSCC): "crowdsourcing" and "case". The search returned 340 articles, most of which were related to crowdsourcing in business. Since the present study focuses on the crowdsourcing activity in the public sector, particularly from the perspective of the government, the study needed to limit the search results to the articles relevant to public administration / policy / planning domain. Thus those articles that were not published in public administration, urban planning or policy journals were eliminated from the search results as the next step. Following that, the results of the search were screened again to select the articles that were directly related to case studies on government crowdsourcing decisions. The criterion to involve case studies in the reviewed articles was preferred, because case studies have the advantage of presenting an in-depth analysis of the issues in question and is useful to explain government crowdsourcing decisions. Therefore, as the final step, those articles that included case studies of government crowdsourcing were marked. The researchers made sure that the selected cases exemplified a variety of application areas for government crowdsourcing. All the selection criteria used in the study finally resulted in 5 main articles to be reviewed. The researchers then assessed each of the selected articles to code evidence on when governments take online crowdsourcing decisions and which factors affect those decisions. The next section describes the characteristics of the evaluated articles and presents findings from their analysis.

FINDINGS

The results of the screening for peer-reviewed articles published in public administration, urban planning, and policy area journals on Web of Science, containing government online crowdsourcing case studies are presented in Table 1 below.

Table 1. Government crowdsourcing case studies

Case 1	Haltofova, B. (2018). Using crowdsourcing to support civic engagement in strategic urban development planning: A case study of Ostrava, Czech Republic. *Journal Of Competitiveness, 10*(2), 85.
Case 2	Hudson, A. (2018). When Does Public Participation Make a Difference? Evidence From Iceland's Crowdsourced Constitution. *Policy & Internet, 10*(2), 185-217.
Case 3	Afzalan, N., & Sanchez, T. (2017). Testing the use of crowdsourced information: Case study of bike-share infrastructure planning in Cincinnati, Ohio. *Urban Planning, 2*(3), 33-44.
Case 4	Christensen, H. S., Karjalainen, M., & Nurminen, L. (2015). Does crowdsourcing legislation increase political legitimacy? The case of Avoin Ministeriö in Finland. *Policy & Internet, 7*(1), 25-45.
Case 5	Lorenzi, D., Chun, S. A., Vaidya, J. S., Shafiq, B., Atluri, V., & Adam, N. R. (2016). Peer: a framework for public engagement in emergency response. *International Journal of E-Planning Research, 4*(3), 29-46.

All of the selected cases are empirical studies on government crowdsourcing. Particularly, they give insights about the government perspective on engaging the citizens. Although the case studies do not directly address factors affecting government online crowdsourcing decisions, in general they imply that governments choose to crowdsource for a variety of purposes. It is also important to note that the selected case studies involve different countries, different levels of government and various public institutions.

The first case study (Case 1) was conducted in Ostrava, Czech Republic. It examines how citizens were engaged by the government crowdsourcing in developing the strategic urban development plan. As the methodology, the case study used data from the online survey of citizens and interviews with the city officials and guarantors of the strategic development plan preparation. According to the findings, crowdsourcing method was preferred in this case, because the city officials aimed to gain the opinion of as many citizens as possible so that the main problems could be identified from the citizens' point of view, involving a broad cross-section of issues. Crowdsourcing helped ensure that the local government plans meet the expectations of the citizens and get widely accepted. In addition, city government wanted to develop a shared vision for the city, and prevent citizens from moving out of the city. As citizens are better aware of the problems occurring in their neighborhood, their experiences and localized knowledge are valuable. Thus, crowdsourcing may be more successfully implemented in local governments and may be more desirable from the perspective of governments, as indicated by the case study in Ostrava. To summarize, crowdsourcing was perceived as highly useful in this case. One of the factors increasing its *perceived usefulness* was the *characteristics of the task*; the process of developing a strategic urban development plan was highly appropriate for widespread public participation and the use of crowdsourcing. As another organizational factor related to crowdsourcing decision, city government possessed the *required resources* to crowdsource in this case. It was able to utilize social media tools and conduct an online survey to get public input. In terms of external influences on the decision to crowdsource, it is apparent that there was *a "crowd" ready to participate*. As local people are generally more interested in the place where they live and are able to comment on major local issues, it may also contribute to the *perceived ease of use* and *usefulness* of crowdsourcing.

Thus, Case 1 highlighted the following factors related to governmental crowdsourcing decisions: perceived usefulness, perceived ease of use, having required resources, availability of the crowd, and characteristics of the task.

The second article, on the other hand, reports on a case study (Case 2) on broader public participation in law making. The study was conducted in Iceland, which is well-known for having the world's first crowdsourced draft constitution. The data used in the study were online records produced in the drafting process, and interviews with participants, including the members of the Constitutional Council. The case study exemplifies how information and communication technology is used to get the public participate directly in the drafting of a constitution. The constitution was drafted using the Constitutional Council's own website, enabling the submission of substantive proposals for the constitution. Also public reflections about the proposals were incorporated to the website through Facebook. As explained in the article, involving the public in drafting the constitution and using crowdsourcing for this purpose is almost uniquely suitable for this kind of process in Iceland case due to its "tiny and homogeneous population, high levels of education, high level of voter turnout (averaging 88 percent since 1946), and a remarkably high level of Internet access, at 96 percent (Kelly et al., 2013)" (Hudson, 2017: 186). This may imply that government *perceived the use of crowdsourcing as easy* in this case, due to the *availability of the crowd* that was interested in political matters and was able to engage with the government online. In addition, government had the *necessary resources* for crowdsourcing, it was able to use Constitutional Council website and facebook very effectively in the drafting process. Besides, since constitution is a legal document that relates to every citizen living in a country, the *task of drafting a constitution* may highly benefit from extensive public participation enabled by crowdsourcing. Thus, *perceived usefulness* of crowdsourcing may also be high in this case due to public value creation and effectiveness of participation activity. Another factor affecting perceived usefulness of crowdsourcing was *political influences*. Involving the public in drafting a constitution was considered to be a political goal and a priority on its own regardless of its quality and quantity, particularly due to its effects on *legitimizing the decisions to be taken*.

In summary, Case 2 suggests that perceived usefulness, perceived ease of use, availability of the crowd, having required resources, characteristics of the tasks and political influences help explain government decisions to crowdsource.

The third case study article (Case 3) highlights Bikeshare planning in the City of Cincinnati as a case study of crowdsourcing. The case demonstrates the use of web-based tools for crowdsourcing by planning organizations to collect ideas and preferences from the public regarding the Bikeshare planning. The article particularly addresses the challenges and concerns of using crowdsourced information, and whether and how those suggestions were integrated into the bike-share plan. It includes data from in- depth semi-structured phone interviews with the two project managers who were involved in using the crowdsourcing tool and creating the Bikesharing plan. Bikesharing planning involves consideration of some issues such as appropriate locations for bike-share stations. The technology used for crowdsourcing in this case was a web-GIS crowdsourcing tool (Shareabouts1), which enabled collecting public's opinion on desired/appropriate locations for bike-share stations.

The case study revealed that "using crowdsourced information in planning processes was related not only to the quality and relevancy of the information, but to other factors such as the organizations' capability of analyzing the information, planner's perceptions of the value of the information, and the planner's attitude towards allocating resources for using the tools and information" (Afzalan and Sanchez, 2017: 42). Another important finding was that, one of the planners used the crowdsourced information mainly to provide support for their interests, in order words, "to legitimize pre-determined elements of a plan" (Afzalan and Sanchez, 2017: 42).

Similar to the earlier case studies described above, Case 3 also emphasized the perceived ease of use and usefulness of crowdsourcing as the major factors related to crowdsourcing decisions. Particularly, new participatory technologies can assist planning and decision-making processes by collecting critical information for policy makers and allowing convenient participation for citizens. However, besides the task characteristics, availability of resources (capacity), and quality and quantity of the crowd affecting these perceptions, the study also demonstrated that incorporating citizens' views to the planning process was an important consideration in crowdsourcing decision because it helps with *legitimizing* the actions to be taken by planners.

Likewise, one of the main factors related to the decision to crowdsource in Case 4 is achieving political *legitimization*, which also explains perceived usefulness of crowdsourcing. The case study investigated in the article shows how crowdsourcing of a legislation through the Finnish website Avoin Ministerio ("Open Ministry") affected the attitudes of the participants towards the political system. The website serves to coordinate crowdsourcing of legislation by delivering online tools for pondering ideas for citizens' initiatives. The developments in citizen attitudes were examined with a two stage-survey of 421 respondents who answered questions about their political and social attitudes, as well as political activities performed. The case study demonstrates that, on the one hand, crowdsourcing motivates the development of innovative proposals for policies by utilizing the combined intellectual resources of citizens. On the other hand, it is also argued that crowdsourcing can assist in generating political legitimacy as it has the potential to develop more positive attitudes toward the political system (Christensen et. al., 2015). Findings from the analyses clearly suggest that crowdsourcing can potentially help improve political legitimacy by creating a more trustworthy decision-making process and enabling the participants to accept the results. Thus, political influence on the crowdsourcing decision is also apparent in Case 4.

Finally, Case 5 relates to one of the major areas of application of crowdsourcing, which is emergency response. Citizen crowdsourcing helps government with achieving better disaster situation awareness and benefitting from the resources provided by citizen volunteers. The focus of the article is the prototype Public Engagement in Emergency Response (PEER) framework, which is an online initiative and mobile crowdsourcing platform for situation reporting and resource volunteering. It aims to share citizen-based disaster situation reports through various communication and social media tools; and providing a platform wherein registered citizens (i.e. volunteers) can share their voluntary services and equipment that they can offer during flood emergencies.

The case study suggests that for crises situations like emergencies, governments' need to cooperate with non-governmental agencies significantly increase. Recently, citizens have become important information sources to assist government emergency response efforts as well, especially through the use of social media. Therefore, perceived usefulness of crowdsourcing tends to be high in emergency response, due to the characteristics of the tasks to be undertaken by government (urgent action and need for information collection and dissemination) and availability of the crowd to assist the government. This case study additionally reveals that crowdsourcing decisions of government agencies might be imitated from similar initiatives in *other public organizations*. Once governments realize the benefits of crowdsourcing in other organizations, they tend to perceive it as more useful and be more likely to adopt this method.

Table 2 below summarizes the variables that emerge out of the reviewed crowdsourcing case studies.

Table 2. Variables that emerge out of the review of the cases

Case 1	Perceived usefulness, perceived ease of use, having required resources, availability of the crowd, characteristics of the task.
Case 2	Perceived usefulness, perceived ease of use, having required resources, availability of the crowd, characteristics of the tasks, political influences (to legitimize the decisions of the policy makers)
Case 3	Perceived usefulness, perceived ease of use, having required resources, availability of the crowd, characteristics of the tasks, political influences (to legitimize the decisions of the policy makers)
Case 4	Perceived usefulness, perceived ease of use, having required resources, availability of the crowd, characteristics of the tasks, political influences (to legitimize the decisions of the policy makers).
Case 5	Perceived usefulness, perceived ease of use, having required resources, availability of the crowd, characteristics of the task, imitating other public organizations

PROPOSED THEORETICAL MODEL AND THE HYPOTHESES

Integrating the key constructs identified by the review of the case studies above, the authors develop a theoretical model of government's decision to adopt online crowdsourcing and propose some hypotheses to be tested in empirical studies. The proposed model depicts crowdsourcing decision as a function of perceived

Figure 1. Theoretical model of government decision to online crowdsource

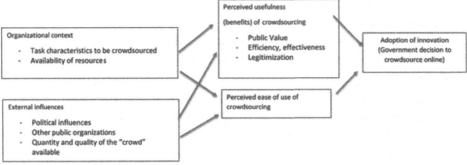

usefulness and perceived ease of use, which are in turn influenced by organizational characteristics and environmental context. The model is shown in Figure 1.

Following Thuan et. al. (2016: 50), the main construct to be explained in this theoretical model, *"government decision to crowdsource"* is conceptualized as "a process that evaluates whether crowdsourcing is an appropriate approach to perform particular organisational tasks". According to the model, decision to crowdsource is influenced by two main factors: perceived usefulness and perceived ease of use.

Perceived usefulness of crowdsourcing is conceptualized as the consideration of the extent to which crowdsourcing can contribute to the realization of governmental goals, including efficiency, effectiveness, political legitimacy, and public value. It is expected that when government perceives crowdsourcing as beneficial for creating public value or realization of other public goals, it may be more likely to adopt this innovation. Therefore, the study posits that:

H1. Perceived usefulness of crowdsourcing will be positively related to government decision to use crowdsourcing.

Based on the model, the authors also propose that perceived usefulness of crowdsourcing is a function of organizational context and environmental context. The authors expect that perceived usefulness of crowdsourcing may depend on *task characteristics* as an organizational factor. For tasks that are clearly defined, have a certain scope, and can be broken down into reasonable parts, expectations regarding the positive outcomes of crowdsourcing decision may be higher, therefore perceived usefulness of crowdsourcing may increase; whereas for more complicated tasks, crowdsourcing may be perceived as less useful due to possible negative anticipations about its consequences. In addition, when there is an urgency regarding the task and public may provide the necessary input for this task faster, more effective and efficient online, crowdsourcing may be perceived as more useful by the government. Based on the model, the study proposes the following hypothesis:

H2. Perceived appropriateness of the government task for crowdsourcing will be positively related to the perceived usefulness of government crowdsourcing.

According to the model, perceived usefulness of crowdsourcing may also be influenced by environmental factors including *availability of the public to participate, political influences, and other public organizations*. Availability of the public that can use the Internet to engage with government is one of the key factors that may affect perceived usefulness of crowdsourcing in environmental context. When the public organization operates in an environment where the demand for public participation tends to be higher and the crowd is already available, such as local governments, asking help from the public and tapping on their opinion, knowledge and expertise online may be perceived as more beneficial.

H3. Availability of the public to provide online input will be positively related to perceived usefulness of government crowdsourcing.

Likewise, when the central and local governments adopt policies that promote public value and support public participation, public agencies may be under more pressure to engage the public. In addition, the need for legitimizing the decisions of the policy makers may motivate them to seek broader public input. Hence, perceived usefulness of crowdsourcing may be higher in these situations. The authors formalize this expectation with the following hypothesis:

H4. Higher political influence will be positively related to perceived usefulness of government crowdsourcing.

As another external influence affecting perceived usefulness of crowdsourcing, other public organizations may also be a role-model in crowdsourcing decisions. According to the proposed model, perceived usefulness of crowdsourcing may be higher when it is observed that other organizations use this technique effectively and benefit from it. Therefore, the study proposes that:

H5. Adoption of crowdsourcing by other organizations in similar cases will be positively related to the perceived usefulness of government crowdsourcing.

According to the model, another factor that affects government crowdsourcing decisions is *perceived ease of use*. Perceived ease of use is conceptualized as the view about whether or not adopting crowdsourcing will be free of effort and risk, and is achievable. In addition to perceiving that online crowdsourcing can be beneficial for their organization, public managers may need to feel that adopting this technique is practical and doable. Therefore, perceived ease of use of online crowdsourcing may affect its adoption decision.

H6. Perceived ease of use of crowdsourcing will be positively related to government decision to use crowdsourcing.

Perceived ease of use, in turn, may be affected by the organizational context and environmental context. It is expected that when public managers perceive that they have available resources to crowdsource online and their tasks are appropriate to open to public participation, they may be less likely to perceive constraints on their innovative behavior and therefore be more likely to decide to adopt this innovation. In line with this, the model hypothesizes the following:

H7. Availability of resources will be positively related to perceived ease of use of crowdsourcing.
H8. Perceived appropriateness of the government task for crowdsourcing will be positively related to the perceived ease of use of government crowdsourcing.

Another factor that may increase the perceived ease of use of crowdsourcing can be the positive experiences of similar organizations that have used crowdsourcing. This expectation is formalized in the following hypothesis:

H9. Adoption of crowdsourcing by other organizations in similar cases will be positively related to the perceived ease of use of government crowdsourcing.

Similarly, higher quantity and quality of public participation available may be perceived as a facilitating condition for crowdsourcing, and may positively affect crowdsourcing adoption. Therefore, it is proposed that:

H10. Higher quality and quantity of public participation to provide online input will be positively related to perceived ease of use of government crowdsourcing.

CONCLUSION

Crowdsourcing online has been popularly utilized especially among business organizations to achieve efficiency and effectiveness goals and to obtain competitive advantage in the market. With the government's increasing interest in using information and communication technologies for a variety of purposes, including achieving efficiency and effectiveness, and creating public value, innovative practices such as online crowdsourcing have also entered into the public administration domain. Accordingly, studies have investigated critical success factors for governmental crowdsourcing, or explored citizen's participation in crowdsourcing activities in case studies. However, factors affecting governmental decision to adopt online crowdsourcing as an innovation have not been sufficiently examined in the literature. While there are studies that explain adoption of open government data or investigate crowdsourcing decisions in the business sector, there is a gap in the literature about the reasons influencing crowdsourcing decisions of government.

In the light of these, this paper aimed to propose a theoretical model that aims to explain government adoption of online crowdsourcing. Based on a review of case studies on governmental crowdsourcing and theories from a variety of disciplines, an integrated theoretical model of factors affecting government crowdsourcing decisions is developed. The model includes perceived usefulness and perceived ease of use as the main factors that affect governmental crowdsourcing decisions. According to the model, these perceptions are, in turn, determined by organizational context (task characteristics to be crowd sourced and availability of resources) and external influences (political influences, other public organizations, and quantity and quality of the crowd available).

The proposed framework may be useful for public managers in crowdsourcing decisions, mainly in determining when to use crowdsourcing and how to use it. The reviewed case studies generally suggest that crowdsourcing has a variety of application areas, and governments may choose to utilize it according to their own priorities, needs and goals. In some instances, crowdsourcing may be preferred for political legitimization purposes, even when the input to be provided by the public may not be useful at all. In other cases, crowdsourcing may be applied because governments may want to benefit from extensive crowd expertise and input, as a support for government efforts in policymaking and implementation. Crowdsourcing may assist governments to accomplish certain goals more efficiently and effectively.

Given that perceived ease of use of crowdsourcing may affect government decisions to utilize or not to utilize this tool, based on the proposed model it can be argued that organizational factors constitute one of the critical components that needs to be carefully considered and managed by public managers. In general, two main organizational factors relate to the feasibility of such projects. First, public managers need to analyze the organizational tasks carefully and decide whether or not they are appropriate for crowdsourcing. Online crowdsourcing can be easier to implement when the tasks are simple and have a well-defined scope. Second, managers need to have a clear understanding about the availability of the organizational resources needed for crowdsourcing. In terms of the technological capacity needed for crowdsourcing as an organizational resource, the review of the case studies shows that governmental websites tend to be one of the most commonly used online crowdsourcing tools, along with the social media tools like facebook. In addition, public agencies need to employ some personnel for the design, implementation, and monitoring of the crowdsourcing activity, as well as for processing its results.

The model also proposes that external influences relate to perceived ease of use of crowdsourcing. In relation to this, availability of the public to be able to give online input is an important external resource to be considered by public managers. In addition, following the best practices of crowdsourcing in other organizations may also be an effective strategy to benefit from this tool.

Besides these managerial implications of the proposed model, the study also aims to make some theoretical contributions. First, by conducting a review of the existing case studies on governmental crowdsourcing decisions with a focus on explaining why governments decide to crowd source, and developing an integrated model as a result, the chapter attempts to fill in the theoretical gap in this area. Second, given that online crowdsourcing is usually considered to be a key practice that can promote creation of public value, understanding factors affecting crowdsourcing decisions in government may also be valuable for theoretically identifying the barriers to achieve public value.

There are also limitations of this study. First, the authors have limited the scope of the systematic review to the governmental crowdsourcing case studies published in the public administration, public policy, and urban planning journals to focus on the public sector perspective. Studies published in other disciplines may offer more insights to the factors related to crowdsourcing decisions in general. Although the present study was able to identify some key constructs related to government crowdsourcing decisions, a larger number of related papers and case studies may contribute to the development of more comprehensive models.

Second, the evidence used in the present study to develop the theoretical model was limited with the content provided by the case study articles, which do not explicitly address the governmental decision making process, thus the evaluation and interpretation of the implications of the case studies by the authors of the present study were required. More specific case studies on governmental crowdsourcing are needed to get a deeper understanding of the factors related to government crowdsourcing decisions.

Third, based on the developed theoretical model, the study proposed some hypotheses to be tested in empirical studies. However, due to data limitations, the present study does not include their testing. Government may choose to crowdsource for a variety of tasks as shown in the crowdsourcing examples. While the study develops an overall model of government online crowdsourcing decision, which of the proposed factors are more influential for what kind of tasks may be tested empirically. Future research may focus on empirically exploring these areas using the hypotheses proposed in this study.

REFERENCES

Afuah, A., & Tucci, C. L. (2012). Crowdsourcing as a solution to distant search. *Academy of Management Review*, *37*(3), 355–375. doi:10.5465/amr.2010.0146

Afzalan, N., Evans-Cowley, J., & Mirzazad-Barijough, M. (2015). From big to little data for natural disaster re- covery: How online and on-the-ground activities are connected. *ISJLP*, *11*(1), 153–180.

Afzalan, N., & Sanchez, T. (2017). Testing the use of crowdsourced information: Case study of bike-share infrastructure planning in Cincinnati, Ohio. *Urban Planning*, *2*(3), 33–44. doi:10.17645/up.v2i3.1013

Ajzen, I. (1991). The theory of planned behavior. *Organizational Behavior and Human Decision Processes*, *50*(2), 179–211. doi:10.1016/0749-5978(91)90020-T

Aladalah, M., Cheung, Y., & Lee, V. C. S. (2016). Delivering public value: Synergistic integration via Gov 2.0. In T. X. Bui, & R. H. Sprague (Eds.), *Proceedings of the 49th Annual Hawaii International Conference on System Sciences (HICSS 2016)*, (pp. 3000-3009). 10.1109/HICSS.2016.376

Asmolov, G. (2015). Vertical crowdsourcing in Russia: Balancing governance of crowds and state–citizen partnership in emergency situations. *Policy and Internet*, *7*(3), 292–318. doi:10.1002/poi3.96

Bandura, A. (1977). Self-efficacy: Toward a unifying theory of behavioral change. *Psychological Review*, *84*(2), 191–215. doi:10.1037/0033-295X.84.2.191 PMID:847061

Bommert, B. (2010). Collaborative innovation in the public sector. *International Public Management Review*, *11*(1), 15–33.

Borst, I. (2010). *Understanding Crowdsourcing: Effects of motivation and rewards on participation and performance in voluntary online activities* (PhD Thesis). Erasmus University Rotterdam.

Brabham, D. C. (2009). Crowdsourcing the public participation process for planning projects. *Planning Theory*, *8*(3), 242–262. doi:10.1177/1473095209104824

Brabham, D. C. (2013). *Crowdsourcing*. Cambridge, MA: MIT Press. doi:10.7551/mitpress/9693.001.0001

Christensen, H. S., Karjalainen, M., & Nurminen, L. (2015). Does crowdsourcing legislation increase political legitimacy? The case of Avoin Ministeriö in Finland. *Policy and Internet*, *7*(1), 25–45. doi:10.1002/poi3.80

Davis, F. D., Bagozzi, R., & Warshaw, P. R. (1989). User Acceptance of Computer Technology. *Management Science*, *35*(8), 982–1003. doi:10.1287/mnsc.35.8.982

DiMaggio, P., & Cohen, J. (2005). Information Inequality and Network Externalities: A Comparative Study of the Diffusion of Television and the Internet. In The Economic Sociology of Capitalism. Princeton, NJ: Princeton University Press.

Doan, A., Ramakrishnan, R., & Halevy, A. Y. (2011). Crowdsourcing systems on the world-wide web. *Communications of the ACM*, *54*(4), 86–96. doi:10.1145/1924421.1924442

Dutil, P. (2015). Crowdsourcing as a new instrument in the government's arsenal: Explorations and considerations. *Canadian Public Administration*, *58*(3), 363–383. doi:10.1111/capa.12134

Ely, D. P. (1993). Conditions that facilitate the implementation of educational technology innovations. *Educational Technology*, *39*(6), 23–27.

Estellés-Arolas, E., & González-Ladrón-de-Guevara, F. (2012). Towards an integrated crowdsourcing definition. *Journal of Information Science*, *38*(2), 189–200. doi:10.1177/0165551512437638

Fishbein, M., & Ajzen, I. (1975). *Belief, attitude, intention, and behavior: An introduction to theory and research*. Reading, MA: Addision-Wesley.

Fitzgerald, C., McCarthy, S., Carton, F., O'Connor, Y., Lynch, L., & Adam, F. (2016). Citizen participation in decision-making: can one make a difference? *Journal of Decision Systems, 25*(sup1), 248-260. doi:10.1080/12460125.2016.1187395

Garcia, A. C. B., Vivacqua, A. S., & Tavares, T. C. (2011). Enabling Crowd Participation in Governmental Decision-Making. *Journal of Universal Computer Science*, *17*(14), 1931–1950.

Haltofova, B. (2018). Using crowdsourcing to support civic engagement in strategic urban development planning: A case study of Ostrava, Czech Republic. *Journal Of Competitiveness*, *10*(2), 85–103. doi:10.7441/joc.2018.02.06

Hansson, K., Muller, M., Aitamurto, T., Light, A., Mazarakis, A., Gupta, N., & Ludwig, T. (2016). Toward a Typology of Participation in Crowd Work. In *Proceedings of the 19th ACM Conference on Computer Supported Cooperative Work and Social Computing Companion* (pp. 515-521). doi: 10.1145/2818052.2855510

Howe, J. P. (2006a). *Crowdsourcing: A Definition*. Retrieved from https://crowdsourcing.typepad.com/cs/2006/06/crowdsourcing_a.html

Howe, J. P. (2006b). The Rise of Crowdsourcing. *Wired, 14*(6). Retrieved from https://www.wired.com/2006/06/crowds/

Howe, J. P. (2008). *Crowdsourcing: Why the Power of the Crowd Is Driving the Future of Business*. New York: Random House Inc.

Hudson, A. (2018). When Does Public Participation Make a Difference? Evidence From Iceland's Crowdsourced Constitution. *Policy and Internet, 10*(2), 185–217. doi:10.1002/poi3.167

Klosterman, R. E. (2013). Lessons learned about planning. *Journal of the American Planning Association, 79*(2), 161–169. doi:10.1080/01944363.2013.882647

Koellinger, P. (2008). The relationship between technology, innovation, and firm performance—Empirical evidence from e-business in Europe. *Research Policy, 37*(8), 1317–1328. doi:10.1016/j.respol.2008.04.024

Lee, S., & Kim, K. (2007). Factors affecting the implementation success of internet-based information systems. *Computers in Human Behavior, 23*(4), 1853–1880. doi:10.1016/j.chb.2005.12.001

Liu, H. K. (2017). Crowdsourcing Government: Lessons from Multiple Disciplines. *Public Administration Review, 77*(5), 656–667. doi:10.1111/puar.12808

Liu, S. M., & Yuan, Q. (2015). The Evolution of Information and Communication Technology in Public Administration. *Public Administration and Development, 35*(2), 140–151. doi:10.1002/pad.1717

Lloret, E., Plaza, L., & Aker, A. (2012). Analyzing the capabilities of crowdsourcing services for text summarization. *Language Resources and Evaluation, 47*(2), 337–369. doi:10.100710579-012-9198-8

Lorenzi, D., Chun, S. A., Vaidya, J. S., Shafiq, B., Atluri, V., & Adam, N. R. (2016). Peer: A framework for public engagement in emergency response. *International Journal of E-Planning Research, 4*(3), 29–46. doi:10.4018/IJEPR.2015070102

Malone, T. W., Laubacher, R., & Dellarocas, C. (2010). The collective intelligence genome. *IEEE Engineering Management Review*, *38*(3), 38–52. doi:10.1109/EMR.2010.5559142

Meijer, A. (2015). E-governance innovation: Barriers and strategies. *Government Information Quarterly*, *32*(2), 198–206. doi:10.1016/j.giq.2015.01.001

Minner, J. (2015). Planning support systems and smart cities. In S. Greetman, J. Ferreira, R. Goodspeed, & J. Stillwell (Eds.), *Planning support systems and smart cities* (pp. 409–425). Cham: Springer International Publishing. doi:10.1007/978-3-319-18368-8_22

Muntés-Mulero, V., Paladini, P., Manzoor, J., Gritti, A., Larriba-Pey, J. L., & Mijnhardt, F. (2013). Crowdsourcing for industrial problems. In J. Nin & D. Villatoro (Eds.), *Citizen in sensor networks* (Vol. 7685, pp. 6–18). Berlin: Springer. doi:10.1007/978-3-642-36074-9_2

Nam, T. (2012). Suggesting frameworks of citizen-sourcing via government 2.0. *Government Information Quarterly*, *29*(1), 12–20. doi:10.1016/j.giq.2011.07.005

Nemec, J., Meričková, B. M., Svidroňová, M. M., & Klimovský, D. (2017). Co-Creation as a Social Innovation in Delivery of Public Services at Local Government Level: The Slovak Experience. In E. Schoburgh & R. Ryan (Eds.), *Handbook of Research on Sub-National Governance and Development* (pp. 281–303). Hershey, PA: IGI Global; doi:10.4018/978-1-5225-1645-3.ch013

Okoli, C., & Schabram, K. (2010). A guide to conducting a systematic literature review of information systems research. *Sprouts: Working Papers on Information Systems*, *10*(26).

Prpić, J., Taeihagh, A., & Melton, J. (2015). The Fundamentals of Policy Crowdsourcing. *Policy and Internet*, *7*(3), 340–361. doi:10.1002/poi3.102

Rogers, E. M. (1995). *Diffusion of innovations* (4th ed.). New York: Free Press.

Saad-Sulonen, J. (2012). The role of the creation and sharing of digital media content in participatory e-planning. *International Journal of E-Planning Research*, *1*(2), 1–22. doi:10.4018/ijepr.2012040101

Seltzer, E., & Mahmoudi, D. (2013). Citizen participation, open innovation, and crowdsourcing challenges and opportunities for planning. *Journal of Planning Literature*, *28*(1), 3–18. doi:10.1177/0885412212469112

Sivarajah, U., Weerakkody, V., Waller, P., Lee, H., Irani, Z., Choi, Y., ... Glikman, Y. (2016). The role of e-participation and open data in evidence-based policy decision making in local government. *Journal of Organizational Computing and Electronic Commerce*, *26*(1-2), 64–79. doi:10.1080/10919392.2015.1125171

Slotterback, C. S. (2011). Planners' perspectives on using technology in participatory processes. *Environment and Planning. B, Planning & Design*, *38*(3), 468–485. doi:10.1068/b36138

Taylor, S., & Todd, P. A. (1995). Understanding information technology usage: A test of competing models. *Information Systems Research*, *6*(2), 144–176. doi:10.1287/isre.6.2.144

Thuan, N. H., Antunes, P., & Johnstone, D. (2016). Factors influencing the decision to crowdsource: A systematic literature review. *Information Systems Frontiers*, *18*(1), 47–68. doi:10.100710796-015-9578-x

Townsend, A. (2013). *Smart cities: Big data, civic hackers, and the quest for a new utopia*. New York: WW Norton & Company.

Triandis, H. C. (1979). Values, attitudes, and interpersonal behavior. In M. M. Page (Ed.), *Nebraska Symposium on Motivation* (Vol. 27, pp. 195-259). Lincoln, NE: University of Nebraska Press.

Twinomurinzi, H., Phahlamohlaka, J., & Byrne, E. (2012). The small group subtlety of using ICT for participatory governance: A South African experience. *Government Information Quarterly*, *29*(2), 203–211. doi:10.1016/j.giq.2011.09.010

Van Duivenboden, H. (2005). Citizen Participation in Public Administration: The Impact of Citizen Oriented Public Services on Government and Citizens. In M. Khosrow-Pour (Ed.), *Practicing E-Government: A Global Perspective* (pp. 415–445). Hershey, PA: IGI Global. doi:10.4018/978-1-59140-637-2.ch018

Wang, H. J., & Lo, J. (2016). Adoption of open government data among government agencies. *Government Information Quarterly*, *33*(1), 80–88. doi:10.1016/j.giq.2015.11.004

Zhao, Y., & Zhu, Q. (2014). Evaluation on crowdsourcing research: Current status and future direction. *Information Systems Frontiers*, *16*(3), 417–434. doi:10.100710796-012-9350-4

Zogaj, S., Bretschneider, U., & Leimeister, J. M. (2014). Managing crowdsourced software testing: A case study based insight on the challenges of a crowdsourcing intermediary. *Journal of Business Economics*, *84*(3), 375–405. doi:10.100711573-014-0721-9

ADDITIONAL READING

Alizadeh, T., Sarkar, S., & Burgoyne, S. (2019). Capturing citizen voice online: Enabling smart participatory local government. *Cities (London, England)*, *95*, 1–10. doi:10.1016/j.cities.2019.102400

El Abdallaoui, H. E. A., El Fazziki, A., Ennaji, F. Z., & Sadgal, M. (2019). An e-government crowdsourcing framework: Suspect investigation and identification. *International Journal of Web Information Systems*, *15*(4), 432–453. doi:10.1108/IJWIS-11-2018-0079

Haltofová, B. (2019). Critical success factors of geocrowdsourcing use in e-government: A case study from the Czech Republic. *Urban Research & Practice*, *§§§*, 1–18. doi:10.1080/17535069.2019.1586990

Hansson, K., & Ludwig, T. (2019). Crowd dynamics: Conflicts, contradictions, and community in crowdsourcing. *Computer Supported Cooperative Work*, *28*(5), 791–794. doi:10.100710606-018-9343-z

Majchrzak, A., & Malhotra, A. (2020). What Is Crowdsourcing for Innovation? In Unleashing the Crowd (pp. 3-46). Palgrave Macmillan. doi:10.1007/978-3-030-25557-2_1

Rexha, B., & Murturi, I. (2019). Applying efficient crowdsourcing techniques for increasing quality and transparency of election processes. *Electronic Government, an International Journal, 15*(1), 107-128.

Shanley, L. A., Parker, A., Schade, S., & Bonn, A. (2019). Policy Perspectives on Citizen Science and Crowdsourcing. *Citizen Science: Theory and Practice*, *4*(1), 1–5. doi:10.5334/cstp.293

Song, Z., Zhang, H., & Dolan, C. (2020). Promoting Disaster Resilience: Operation Mechanisms and Self-organizing Processes of Crowdsourcing. *Sustainability*, *12*(5), 1–17. doi:10.3390u12051862

KEY TERMS AND DEFINITIONS

Crowd: Anyone who is aware of the task to be crowd sourced and who can use the Internet to contribute to such a goal.

Crowdsourcing: The activity of getting information or input for a task or a project in various sectors from a large and relatively open group of internet users.

Innovation: A new thing, method or technique.

Legitimization: Making something acceptable to the public.

Perceived Ease of Use: An individual's belief that using a tool, a method, or a technique will be easy and simple.

Perceived Usefulness: An individual's belief that using a tool, a method, or a technique will bring some benefits.

Technology Adoption: The behavior of accepting, owning and using a new technology.

Chapter 9
Mudamos:
A Mobile App to Ignite an Integrated Engagement Framework

Marco Konopacki
Institute for Technology and Society, Brazil

Debora Albu
Institute for Technology and Society, Brazil

Diego Cerqueira
https://orcid.org/0000-0003-4861-3394
Institute for Technology and Society, Brazil

Thayane Guimarães Tavares
Institute for Technology and Society, Brazil

ABSTRACT

Brazil's Constitution established a few means of direct democracy including the possibility of any citizen to propose a draft bill at a legislative house given the support of a minimum of citizens expressed by their signature. Until today, citizens' initiative bills' signatures are paper based, which is not only costly, but also poses transparency and safety issues. Considering these challenges, the Institute for Technology and Society developed a mobile app called "Mudamos" ("We Change") to prove it is possible to sign such bills electronically. However, despite its potential for changing citizen participation, technology by itself does not promote political and cultural changes. Thus, Mudamos became an integrated engagement framework, including the free application and also an offline legal draft-a-thon and advocacy for institutional change. In this chapter, the authors present this framework, connecting cutting-edge digital innovation on electronic signatures with social innovative methodologies, highlighting the importance of adopting a holistic approach to institutional change.

DOI: 10.4018/978-1-7998-1526-6.ch009

INTRODUCTION

Brazil's Constitution established a few means of direct democracy, including the possibility of any citizen propose a draft bill at a legislative house at the municipal, state or federal level, given the support of a minimum of citizens explicit through their signature (Teixeira, 2008; Konopacki and Itagiba, 2017). Until today, popular initiative bills' signatures are paper-based, which is not only costly, but also presents problems connected to transparency and safety principles. In fact, there are no citizens' bills approved at the national level due to the verification barrier. All the projects proposed at this level which collected the minimum signatures necessary had to be adopted by a member of parliament.

To change this scenario, the Institute for Technology and Society (ITS) created "Mudamos" (We Change) in 2017 to reduce the high costs of creating paper-based petitions by offering a verifiable online mechanism for the creation and signing of citizen petitions, offering a robust means of participation that, in turn, should help to raise citizens' degree of trust in political institutions and contribute to the construction of participatory rules and norms (Avritzer, 2019).

"Mudamos" is a mobile application that enables Brazil's citizens to participate in lawmaking by proposing their own bills and signing onto one another's proposals using verified electronic signatures, reducing the high costs of creating paper-based petitions. Any citizen with a smartphone (Android or iOS) can download the app and register with their electoral ID, name and address, information which "Mudamos" keeps secure and verifies with Brazil's Electoral Court. The app issues what is known as a cryptographic key pair, (Public-key cryptography, 2020) a small piece of code used for verification. One half of the key is stored on the user's phone and the other with Mudamos, which makes it possible to authenticate a person's signature. This way, members of the public can draft and sign petitions in a way that is verifiable and secure. With 700,000 people downloading and signing up in the first year of this crowdlaw initiative, "Mudamos" could be the linchpin to enabling crowdsourcing drafting of legislation and unlocking the power of direct democracy in practice.

Despite the fact that apps such as Change.org or Avaaz.org have increased their number of petitions and signatures in the last few years, they face problems implementing the demands in effective institutional impact (Aragón et al, 2018). It means that engaging people through digital tools was not enough to promote change without offline coordination efforts. In the same way, as time passed after Mudamos' launch, the team realized that lowering participatory costs using an electronic channel was not the only key to strengthen and to sustain engagement. How could Mudamos be revamped into an effective tool to enhance civic participation? Was the creation of a civic technology enough to deeply transform participation dynamics in Brazil?

The authors argue that for real institutional changes, it would be necessary to articulate an integrated engagement framework (Eccles, 2016) combining 1) electronic participation means with 2) offline deliberative spaces and 3) an advocacy strategy. These three pillars that structure Mudamos as a project were developed in the past three years based on the hypothesis that online participation tools were a necessary but insufficient measure to fully transform the dynamics of civic participation in the country .

This chapter takes the concept of crowdlaw as a starting point (Noveck, 2017) and it explores the literature around it to position the development of "Mudamos" as a project and a mobile application. It goes on to describe the regulatory environment for citizens' initiative draft bills and explains why it has not been implemented as a "de facto" right in Brazil. The following section describes Mudamos' team approach to creating an engagement framework which comprises three strategies: 1) a civic technology tool taking the format of a mobile application, 2) an offline methodology to collaboratively draft bills ("Draft-a-thon"), and 3) an advocacy strategy to make institutional changes based on crowdlaw principles at legislative houses. Finally, in the concluding remarks, the research highlights the need to combine technology, especially in the form of digital channels, with other mechanisms that can promote truly enduring changes to produce social and institutional impact.

Literature Review: Crowdsourcing and Crowdlaw

e-Government is understood as the study of the development and delivery of government services (for citizens and businesses) through digital platforms (Gupta et al., 2016, p. 161). Within this field of research, the theoretical and empirical disputes are about the topics that deserve attention and the methodologies used for investigation (Dutton, 1992).

The main topic of interest for the e-government debate is the diffusion of e-government practices (Zhang et al., 2014, p. 3; Gupta et al., 2016; Titah & Barki, 2006, p. 23). In other words, understanding how certain practices of governments can contribute to its branches and institutions to digitize their services. From the citizens' perspective, it means how they are encouraged to use online public services provided by public institutions (Tomkova, 2009).

The term crowdsourcing was first mentioned by the American journalist Jeff Howe in his article published in Wired magazine in June 2006 under the title The Rise of Crowdsourcing. (Howe, 2006) In his article, Howe defines crowdsourcing as "the act of a company or institution taking a function once performed by employees and outsourcing it to an undefined (and generally large) network of people in the form of an open call."

Figure 1. Main screenshots of app "Mudamos"

In subsequent years, the use of the term crowdsourcing gained popularity. By 2012, there were around 209 indexed articles that included the term crowdsourcing among their keywords. (Estellés-arolas; González-ladrón-guevara, 2012) In their article "Towards an integrated crowdsourcing definition", Enrique Estellés-Arolas and Fernando González-Ladrón-De-Guevara systematized the content produced on crowdsourcing until 2012 and sought to establish a comprehensive concept of the phenomenon based on the different works published so far. The authors based analysis in three dimensions: the crowd, the initiator, and the process. There is no explicit theoretical foundation for the decision to consider these three dimensions of analysis. According to the authors, these three dimensions emerged based on the review of the collected material. After the systematization of the formulations discussed in the works found, the authors came up with the following proposal of definition of the concept:

Crowdsourcing is a type of participative online activity in which an individual, an institution, a non-profit organization, or a company proposes to a group of individuals of varying knowledge, heterogeneity, and number, via a flexible open call, the voluntary undertaking of a task. The undertaking of the task, variable complexity and modularity, and in which the crowd should participate bringing their work, money, knowledge and / or experience, always entails mutual benefit. The user will receive the satisfaction of a given type of need, be it economic, social recognition, self-esteem, or the development of individual skills, while the crowdsourcer will obtain and utilize to their advantage what the user has brought to the venture, whose form will depend on the type of activity undertaken. (Estellés-arolas; González-ladrón-guevara, 2012, p. 197)

This definition establishes eight objective criteria for assessing whether an initiative can be considered crowdsourcing or not, considering the presence of all of them in the initiative: 1) a clear definition of the crowd; 2) the definition of tasks and their objectives; 3) the reward given for the established work; 4) the existence of a convener of the case; 5) the establishment of the gains achieved by the convener; 6) the process is online and participatory; 7) the call is open, even with variable coverage; and 8) use the internet.

Crowdsourcing differs from other forms of participation by its problem solving orientation. For Daren Brabham, crowdsourcing allows organizations in the face of problems that they cannot solve internally, transferring the task to a community, broadening the contribution base to think of a solution. The Internet allows the base of contributions to expand exponentially, neither geographically nor temporally restricted. By broadening its workers base, it also broadens the diversity of skills, competencies and ideas to tackle the problem, allowing not only expert input on a topic, but also ideas from other non-expert audiences, who can somehow contribute to the solution. (Brabham, 2013, p. 18-20) þ

On the other hand, crowdlaw is a concept that has been gaining popularity since the mid-2010s, during which it has been used in some studies related to citizen participation in lawmaking and public policymaking. Roughly speaking, crowdlaw is the result of the combination of the words crowdsourcing and law. The organization directed by Beth Simone Noveck, The Governance Lab defines crowdlaw as "the practice of using technology to tap the intelligence and expertise of the public in order to improve the quality of lawmaking" (The Govlab, 2019a)

For crowdlaw scholars, it is a method to modernize public services and policymaking through the intensive application of new technologies, such as big data and artificial intelligence, to bring into the State citizens expertises. (Noveck, 2015, 2017) It is a mix of participatory policies and e-government approaches which put the State as the primary actor in designing and implementing citizens-first public services.

From this perspective, the State has an interest in providing a channel for open participation to improve the provision of public services. The key element for thinking about this relationship is crowdsourcing. In crowdsourcing definition, the State is the the calling organization which works in balance with the community of citizens that contributes to execution of tasks or services.

Citizens' Initiative Draft Bills in Brazil

Since the Federal Constitution was approved in 1988, Brazil has had a law which allows citizens to propose draft bills to legislative houses at all levels - municipal, state, and federal - once the proposal meet the requisite threshold for signatures. At the federal level, a draft bill needs a minimum of 1% of registered voters support - which in absolute terms translates into 1.5 million verified signatures - to be presented to the Lower House. At the majority of states, the 1% percentage is also the rule, and, at the municipal level, the general rule is to have 5% of voters supporting the bill[1]. The logistical barriers to collect and validate these signatures, together with voters' identifications and addresses, are the greatest obstacles in this process, on top of signature verification (Teixeira, 2008). As a result, to date no citizen initiated bill has ever been approved as such due to these challenges: it is very difficult to collect and manually verify more than one million and a half signatures.

It is relevant to highlight that citizens initiative draft bills are binding, meaning it is mandatory that they be received and discussed by the legislative house once the threshold is reached. After being received, they go through two rounds of discussion by members of parliament, first, at special commissions and, then, in plenary sessions, the same way parliamentary initiative bills are. During these phases, their texts can be altered by parliamentarians, which can be a weak point from a civic participation perspective. In that sense, they are not comparable to petitions as they are not a manifestation of civic participation and discontent, but carry the weight to entry into force as regular legislation in the country.

In a few instances, interested politicians have "adopted" the draft bill and presented it as if they authored it, eliminating the need for verifying the signatures (Konopacki & Itagiba, 2017). In other words, congressmen act as proxies for citizens to claim their direct participation rights. Some of the projects that were "adopted" since 1988 include Law 8930 of 1994 (on femicide), Law 9840 of 1999 (to fight electoral corruption), Law 11124 of 2005 (that creates a National Fund for Social Housing) and Complementary Law 135 of 2010 (known as the "Clean Slate Law").

Brazil's Federal Constitution was developed and written before the democratization of the Internet and the digital world people are immersed in. For that reason, signatures and information for citizens' draft bills have been collected in the possible format at the time, meaning through paper and pen. Brazil is a country with a huge population and a large territory, which results in even more difficulties to collect this support. Besides this, since the signatures are paper-based there is a logistical barrier both in their collection as well as in validation of the information needed (names, addresses, voters' identification numbers) (Teixeira, 2008).

Considering the progress of technologies of communications and information (ICTs), the legal adaptation for the collection of such signatures through the Internet is ever more necessary. Political behaviors of both citizens and politicians has been changing rapidly together with new technological, social and political contexts: political participation is not exclusive of the offline world and institutions become increasingly more open to Internet usage (Almeida, 2015). Currently, it is possible to collect signatures throughout Brazil and verify them automatically. Digital signatures already have their relevance recognized and used in common civic procedures through the Law MPV2200 from 2001 and in legal acts as established by Law 11419/064[2].

Nonetheless, only turning the signature process into electronic does not foster public engagement by itself. Mudamos caught the attention of the public and it was massively downloaded during the following months after its launch. The app was downloaded more than 700k times and one draft bill proposal reached more than 250k of signatures. However, the mobilization chilled after the early hype. A literacy barrier on how people could propose new draft bills and communicate them was identified. In addition, people were sending more draft bills proposals than the core team could analyse, which made the time to upload a new draft bill on the platform very long. In face of these challenges, Mudamos' team developed an integrated engagement framework, which this chapter turns now to describe.

An Integrated Engagement Framework

In a strategic turning point, Mudamos was combined with other side initiatives aiming to overcome challenges faced after the project launch. The first challenge - to make the draft bill analyses' time shorter - Mudamos created a *volunteer program*. The second challenge was tackled with the creation of a *legislative draft-a-thon*, supporting citizens to propose new draft bills and to learn how to engage people on their campaigns. The final challenge is to fully integrate the electronic signature format into municipal, state and the federal constitution, institutionalizing the transformation brought by technology, which allows citizens access to this right. To tackle this challenge, the team designed an *advocacy strategy* for electronic signatures.

Volunteers Program

Mudamos not only allows users to sign in support of a bill but also to suggest new draft bills for signature collection. To propose a draft bill, the user needs to answer the following questions: 1) draft bill name; 2) draft bill content; 3) which level the bill addresses (national, state, municipality); 4) video description (optional); 5) whether there is a law related to this draft bill (yes or no).

Since Mudamos has been released, the platform has received more than 8,000 draft bill proposals. In its early days, the Mudamos legal team consisted of a single specialist who was responsible for analyzing the large number of proposals. To

Figure 2. Screenshot of Mudamos draft bill proposal system

Olá, você está na área de proposição de projetos de lei de iniciativa popular do aplicativo Mudamos.
Siga os próximos passos para enviar seu projeto :)

 carrega em ENTER

address the volume, Mudamos' creators have designed a volunteer program. Since January 2018, ITS uses crowdsourcing to engage young lawyers to assist in the analysis of the proposals.

The volunteer program had 63 applications, mostly by young lawyers. Many were from Rio de Janeiro, but all regions were represented in the applicant pool. After an evaluation process, Mudamos selected 26 volunteers who attended a course supported by ITS. The course was delivered as a webinar (online) in March 2018 and covered topics such as the legislative process, draft bill proposers communication protocol and sharing experiences.

Before draft bills are published in the platform, the Mudamos volunteer legal team performs a legal analysis to verify whether the draft bill has all the constitutional requirements to be framed as a formal petition. If it satisfies the constitutional requirements, the bill is uploaded on the platform and published for signature-gathering immediately. If it has not, the bill's author receives a feedback report based on the analysis recommending changes or explaining why the proposal cannot be accepted as a citizens' initiative bill.

Since the beginning of the volunteer program, more than 600 proposals were analysed and from those, 30 bills were considered constitutional and uploaded on the platform. Due to the success of the program, the team has initiated a new cycle of volunteers in 2019 and they have created a new technological system for distributing and registering the analyses, automating the process and reducing the time spent on this activity even more.

Legislative Draft-a-Thon

Eighteen months after its initial release, Mudamos improved not only its software platform but also its workflow. As new features were added it was necessary to create procedures to face new project challenges. When Mudamos began to accept new draft bill proposals, it quickly became clear that people did not know how to format their petitions as a formal bill.

There were many good ideas that were not properly formatted as a draft bill proposal, providing more strain on the volunteer legal team to get the bill into shape. To overcome this issue, Mudamos team created a side project called *Virada Legislativa* (legal draft-a-thon). The Virada Legislativa is a one-day in-person event to develop draft bills collectively -- a draft-a-thon -- addressing a single issue and within a timeframe. To assist during the process, there are trained and volunteer facilitators: one group of experts in law - Law students, public servers at legislative houses, lawyers - and a second group of experts on the topic that is being discussed in that draft-a-thon (e.g.: for urban mobility, there were urbanists, engineers and researchers at the event).

The draft-a-thon is divided into six stages. The first stage consists of a multi-stakeholder panel to focus on establishing the basis for the debate and an introductory reflection moment for participants to recognize each other and their ideas in the group. The next stage is a fishbowl conversation with the audience, including the first stage panelists, where participants will give their input on which are the main problems related to the topic. (Fishbowl conversation, 2019). In the third stage, facilitators start to cluster the discussion into thematic axes that will be the basis for forming groups to draft the bills in the following sections.

The following stage is directed to the actual drafting of the proposals and based on the first reflections: addressing the draft bill main objective, writing down general definitions and, finally, drafting bill articles. At this point the group is divided into the thematic areas previously defined as the result of the three first stages of the draft-a-thon. All working groups are supported by the facilitators (both specialists in the issue addressed as well as legal experts) to draft the bills.

The last two stages are the "test" of the bill and its publication on *Mudamos*. For the "test," the group chooses a representative to present the proposal to a stand formed by parliamentarians who play the role of consultants to improve the bill. These are actual parliamentarians who will in the future discuss the bill. This stage was thought as a moment to narrow the gap between citizens and lawmakers, creating empathy from both sides, so they understand power, responsibilities and duties. The final stage is the publication of the draft bill on Mudamos and its subsequent signature collection.

Figure 3. Draft-a-thon phases and methodology

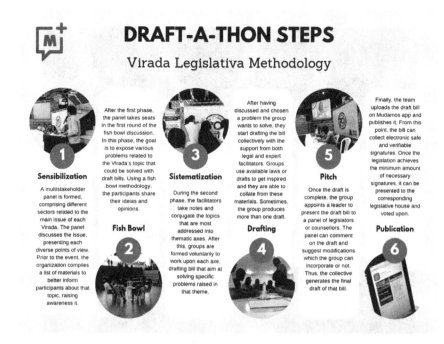

"Virada Legislativa" is ruled by three principles:

a) **Multi-stakeholder approach**: the more diverse the sectors participating in the activity, the stronger will be the proposal drafted, as it will take into account various points of view and interests.

b) **Collaboration**: all the participants co-create the proposals, exchanging ideas and trying to reach consensus, with active listening and non-violent communication. This is mediated by group facilitators previously trained by Mudamos's team.

c) **Open call and online/offline interaction:** the activity is open to the participation of all citizens, without any kind of selection. Despite the draft-a-thon taking place in person, it is connected to the online space through tools such as live streaming, commenting and suggesting on digital platforms to add more input the the in-person deliberative space.

The first "Virada Legislativa" took place in João Pessoa, a city in Paraíba in the Northeast of Brazil, and addressed urban mobility, having the collaboration of civil society organizations, academia, and local public administration. More than 100 people participated in the activity as well as 20 group facilitators and 15 city councillors. The participants collectively drafted five proposals on issues such as

Figure 4. Picture from the first Virada Legislativa held in the city of João Pessoa, state of Paraíba. Credit: Yasmin Tainá.

transport integration, sidewalk standardization and open data on the transport system. Another "Virada Legislativa" took place in Rio de Janeiro on entrepreneurship. In this legal draft-a-thon, one draft bill was developed on decreasing the highly bureaucratic procedures for opening a company in the country. Besides these "Viradas," various workshops on this methodology have been conducted, aiming at multiplying its applicability.

To date, there have been three "Virada Legislativa" events with 200 participants leading to new 11 draft bills on the platform which altogether have received 6014 signatures. Despite the low levels of subsequent participation on the app, the live events had a significant impact. One of the most important outputs of the "Viradas Legislativas" is to join city councillors and citizens together to collaborate around law making. The city of João Pessoa, where two of Viradas Legislativas took place, changed their participatory culture from ordinary consultation process to a real collaboration between citizens and parliamentarians.

Figure 5. fishbowl conversation with a counselor talking at the same level of citizens. Credit: Yasmin Tainá.

Advocacy for Institutional and Cultural Changes

With the massive adoption within a few months after its launch, Mudamos has been leading, not only a technological turn in politics, but also fostering institutional and cultural changes. Perhaps the most significant impact of the program is the transformation of political life from a large closed door to a now more participatory process. Before Mudamos, Brazilian citizens were generally not aware they could propose draft bills. After Mudamos was released they have been excited with the "new" institutional mechanism available to them to influence politics. At the same time, politicians have started to pay attention to Mudamos and its ability to make easier the signature gathering process for citizens' initiatives draft bills.

At the national level, in late 2016 and throughout 2017, Congress was discussing the issue of political reform. In April 2017, a few days after Mudamos' first release, ITS Rio was invited to discuss and present Mudamos to the congressional Political Reform Committee, led by its speaker congressman Vicente Cândido. That hearing led to draft bill 7574 of 2017, which would update the citizens' initiative law to recognize the new electronic mechanism.[3] The bill is still under negotiation in the National Lower House and is ready for a floor vote. In addition to this legislative reform, Mudamos supported the efforts of the representatives to facilitate the use of electronic signatures for civic participation by addressing congressional internal rules and procedures. In 2017, Congressman Alessandro Molon presented two proposals to make Congress ready to receive electronic signatures for citizens' initiative draft bills.[4]

Taking into account the difficulty to make changes at the national level, the Mudamos team also directed efforts on local changes. In João Pessoa, capital of Paraíba state, municipal law 13041 (2015) regulates the use of electronic signatures in petitions.[5] However, since its approval, there have been no adequate technical tools to realize this law. Recognizing Mudamos as a cheap and accessible technical option, on May 9, 2017, at a public hearing held by the city council, Mudamos was designated the official channel to present citizens' initiative draft bills.

Besides Mudamos being widely recognized by the general public, some public servants and representatives have also turned it into the main channel for draft bill proposing in other states. With their experience in João Pessoa serving as inspiration for institutional change, Mudamos team created a legal framework to support both legislative houses of Congress with updating their procedures. This legal framework is a collection of documents that can be used by representatives and public servants as a template to create new norms to allow electronic signatures to be accepted in their legislative houses. It is equally useful for state and local legislatures as well as the federal level. Following this framework, the City of Divinópolis in Minas Gerais institutionalized Mudamos through a memorandum of understanding where ITS supports the efforts of their legislative houses to update their norms and procedures to be prepared to accept electronic signatures.

In addition to changing norms and practices, Mudamos has also strengthened Brazilian political culture and literacy on collaborative law making. Since the launch of Mudamos, people have demonstrated they have a strong will to participate and good ideas to propose. Without any experience with political participation, the knowledge of how to be active in political processes is underdeveloped. The lack of experience combined with the arcane and legalistic nature of the lawmaking process, which is very jargon-filled and detached from citizens' everyday reality, have given rise to anticipated challenges of needing to "translate" between the needs and desires of ordinary citizens and the formalistic demands of the legislative process. Through its integrated engagement framework, Mudamos is contributing to overcome these challenges using a multidimensional approach to tackle structural malfunctions in Brazilian participatory system.

CONCLUSION

In the early days of e-participation studies, electronic tools that had been created were considered the very innovative aspect towards the construction of a digital democracy. Experiments such as systems for online deliberation and teledemocracy were taken as an opportunity to "fix democracies old problems" and engage citizens to participate in politics. The main expectation was that those new tools would be able to transform how people interact with their governments and through new ways of communication, take them closer to their representatives, making it easier to watch, and interact with them.

Nonetheless, after some experiments using electronic tools for e-public participation, especially on e-consultations, some studies described that often this kind of tool was being used more as an instrument of governments to increase their political capital, than a way to create real institutional changes. Although people had new channels to express themselves politically, "outcomes of e-consultation initiatives have been poorly and arbitrarily integrated in the respective policies they intended to inform. Their inclusion has remained contingent on the political will and discretion of the political actors" (Tomkova, 2009).

Focusing only on such tools for e-participation and disconsidering that they are immersed in complex political systems could lead to misdiagnosis on how digital tools can change political structures and transform institutions. Old democracies problems such as political apathy, lack of literacy and power games are not only solved by opening them through electronic tools. Otherwise it should combine electronic tools with other efforts tackling those problems in a holistic way, which is not only using electronic tools, but also support by classical deliberative practices.

In this chapter, the researchers described the project Mudamos and its path from a singleton e-participation tool to an integrated engagement framework, which takes into account that there are users limitations and institutional challenges, which are now tackled fostering various approaches and methodology in order to produce social and political impact. Mudamos' case informs that this holistic approach to democracies challenges is the strongest one and should be considered in other contexts.

REFERENCES

Almeida, G. de A. (2015). Marco Civil da Internet – Antecedentes, Formulação Colaborativa e Resultados Alcançados. In Marco Civil da Internet Análise Jurídica Sob uma Perspectiva Empresarial. São Paulo: Quartier Latin.

Alsina, V., & Martí, J. L. (2018). The Birth of the CrowdLaw Movement: Tech-Based Citizen Participation, Legitimacy and the Quality of Lawmaking. *Analyse & Kritik, 40*(2), 337–358. doi:10.1515/auk-2018-0019

Aragón, P., Sáez-Trumper, D., Redi, M., Hale, S., Gómez, V., & Kaltenbrunner, A. (2018, June). Online Petitioning Through Data Exploration and What We Found There: A Dataset of Petitions from Avaaz. org. *Twelfth International AAAI Conference on Web and Social Media.*

Avritzer, L. (2009). *Participatory institutions in democratic Brazil.* Woodrow Wilson Center Press.

Avritzer, L. (2019). *O pêndulo da democracia.* São Paulo: Todavia.

Brabham, D. C. (2013). *Crowdsourcing.* Cambridge: MIT Press. doi:10.7551/mitpress/9693.001.0001

Brabham, D. C. (2015). *Crowdsourcing in the Public Sector.* Georgetown University Press.

Dutton, W. H. (1992). *Political Science Research on Teledemocracy. n. 1977.* Academic Press.

Eccles, J. S. (2016). Engagement: Where to next? *Learning and Instruction, 43,* 71–75. doi:10.1016/j.learninstruc.2016.02.003

Estellés-Arolas, E., & González-Ladrón-de-Guevara, F. (2012). Towards an integrated crowdsourcing definition. *Journal of Information Science, 38*(2), 189–200. doi:10.1177/0165551512437638

Fishbowl (conversation). (2019). Retrieved from https://en.wikipedia.org/wiki/Fishbowl_(conversation)

Gupta, K. P., Singh, S., & Bhaskar, P. (2006). *Citizen adoption of e-government: a literature review and conceptual framework.* Academic Press.

Howe, J. (2006). The rise of crowdsourcing. *Wired Magazine, 14,* 1-4.

Konopacki, M., & Itagiba, G. (2017). *Relatório Projetos de Lei de Iniciativa Popular no Brasil.* Instituto de Tecnologia e Sociedade. Available at: https://itsrio.org/wp-content/uploads/2017/08/relatorio-plips-l_final.pdf

Noveck, B. (2010). The Single Point of Failure. In Open Government: Collaboration, Transparency, and Participation in Practice. O'Reilly Media Inc.

Noveck, B. S. (2015). *Smart citizens, smarter state: The technologies of expertise and the future of governing.* Harvard University Press. doi:10.4159/9780674915435

Noveck, B. S. (2017). Crowdlaw. *Collective Intelligence and Lawmaking.*, *40*(2), 359–380.

Public-key cryptography. (2020). Retrieved from https://en.wikipedia.org/wiki/Public-key_cryptography

Teixeira, L. A. (2008). *A Iniciativa Popular de Lei no Contexto do Processo Legislativo: Problemas, Limites e alternativas. 2008: Monografia para Curso de Especialização. Câmara dos Deputados: Centro de Formação, Treinamento e Aperfeiçoamento.* Available at: http://bd.camara.gov.br/bd/handle/bdcamara/10190?show=full

The Govlab. (2019a). *Crowdlaw - The Governance Lab.* Available at http://www.thegovlab.org/project-crowdlaw.html

The Govlab. (2019b). *Crowdlaw for Congress - The Governance Lab.* Available at https://congress.crowd.law/

Titah, R., & Barki, H. (2006). E-government adoption and acceptance: A literature review. *International Journal of Electronic Government Research*, *2*(3), 23–57. doi:10.4018/jegr.2006070102

Tomkova, J. (2009). E-consultations: New tools for civic engagement or facades for political correctness. *European Journal of ePractice, 7*, 45-55.

Zhang, H., Xu, X., & Xiao, J. (2014). Diffusion of e-government: A literature review and directions for future directions. *Government Information Quarterly*, *31*(4), 631–636. doi:10.1016/j.giq.2013.10.013

ENDNOTES

[1] Since Brazil is a federation, each state and municipality can decide upon a series of matters, including the percentage of voters' support for a popular initiative draft bill. For example, João Pessoa, a municipality in the Northeast of Brazil, has changed its municipal constitution and the threshold for presenting such bills is of only 0,5% in contrast with most municipalities in the country, where the minimum is of 5%.

[2] Law No. 11419 of 2006 regulates the digitalization of the legal process.

3   See https://www2.camara.leg.br/atividade-legislativa/discursos-e-notas-taquigraficas/discursos-em-destaque/reforma-politica-1/relatorio-parcial-1-da-reforma-politica.

4   See http://www.camara.gov.br/proposicoesWeb/prop_mostrarintegra;jsession id=8BE310908A0DBAC8152C6D999F3CA310.proposicoesWebExterno2?c odteor=1540588&filename=INC+3228/2017 and http://www.camara.gov.br/proposicoesWeb/prop_mostrarintegra?codteor=1542795.

5   See https://leismunicipais.com.br/a/pb/j/joao-pessoa/lei-ordinaria/2015/1305/13041/lei-ordinaria-n-13041-2015-disciplina-a-iniciativa-popular-de-leis-a-que-se-refere-o-art-31-da-lei-organica-do-municipio-de-joao-pessoa?q=13041.

APPENDIX

Screenshots of Main Canvas of Mudamos App

Figure 6.

Figure 7.

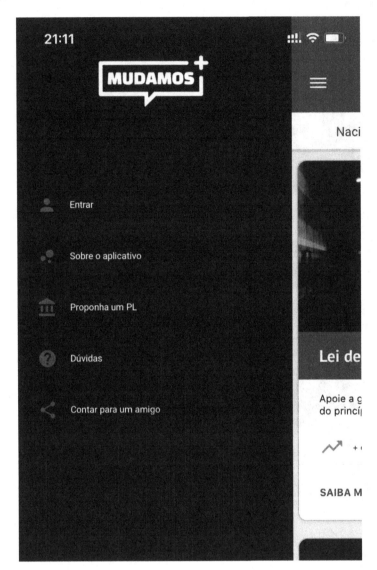

Figure 8.

**Oi! Com o Mudamos, você pode assinar
projetos de lei de iniciativa popular e
transformar a sua cidade, estado ou o país. E
tudo o que você precisa é do seu celular! :)**

O que é um projeto de lei de iniciativa popular?   ∨

Por que criar um aplicativo para propor leis de
iniciativa popular?   ∨

As assinaturas feitas por meio do aplicativo são
válidas?   ∨

Eu só posso assinar um projeto de lei usando o
aplicativo?   ∨

Por que pedimos seu CPF e Título de Eleitor?   ∨

Figure 9.

Figure 10.

Assinaturas: 154.293 Meta: 1.700.000

Ajude a acabar com a compra de apoio político no Brasil

Chapter 10
Smart Place Making Through Digital Communication and Citizen Engagement:
London and Madrid

Angel Bartolomé Muñoz de Luna

https://orcid.org/0000-0001-7056-8855
CEU San Pablo University, Spain

Olga Kolotouchkina

https://orcid.org/0000-0002-8348-8544
CEU San Pablo University, Spain

ABSTRACT

The disruptive growth of new information technologies is transforming the dynamics of citizen communication and engagement in the urban context. In order to create new, smart, inclusive, and transparent urban environments, the city governments of London and Madrid have implemented a series of innovative digital applications and citizen communication channels. Through a case study approach, this research assesses the best practices in the field of digital communication and citizen engagement implemented by London and Madrid, with a particular focus on the profile, content, and functions of these new channels. The results of this research are intended to identify relevant new dynamics of interaction and value co-creation for cities and their residents.

DOI: 10.4018/978-1-7998-1526-6.ch010

INTRODUCTION

The impressive growth of information and communication technologies (ICTs) since the end of the 20th century, is transforming the metabolism of modern cities. Places, people and objects have been linked together through digital flows of data and content that create new social dynamics and shape new paradigm of urban place making (Castells, 1996; Sassen, 2011). Virtual representation and digital environments have become essential policy-making tools for an effective engagement of urban stake-holders around key values, culture and challenges of urban places (Govers, 2015).

Although the impact of urban innovation associated with ICTs is quite uneven throughout the world, the smart city model is an increasingly common reference in professional forums and academic debates about the challenges of prosperous and sustainable urban development (Albino, Berardi, & Dangelico, 2015). Smart cities are beginning to be identified as an exceptional way of improving the efficiency of urban services and the quality of life of city residents, reducing environmental impact and establishing a new model for relations between all urban stakeholders (Caragliu, Del Bo, & Nijkamp, 2011; Giffinger et al, 2007).

In the context of this new smart urban logic, the role of citizens as key actors of urban development is getting greater attention. The concept of citizen participation is being revisited, fostering the paradigm of active citizen engagement and participatory democracy as key premises of smart place making. In parallel to the emergence of new formulas allowing greater legitimation of the political authorities, meaningful efforts from policy-makers are oriented towards the idea of participatory democracy, developing new procedures to incorporate citizens into public decisions on social transformation strategies (Alguacil, 2006). Furthermore, citizen engagement takes on particular importance in participatory budgeting, when citizens are not only encouraged to propose the development of specific projects and social initiatives in their cities, but also to cast their vote to carry them out.

The impact of ICTs on the everyday reality of a city and some new formats of interpersonal communication emerging between urban residents can be seen in the new dynamics of social cooperation and citizen mobilisation (Rheingold, 2002). The founding of social activist groups operating both in physical and virtual urban environments is increasingly common. The emergence of smart communities, together with the ease of access to urban data and content through the many municipal apps, contribute to consolidating collective intelligence, citizen innovation and a more active, responsible attitude to urban development (Capdevila & Zarlenga, 2015; Albino et al., 2015).

Alongside the growing public activism on social networks, the smart urban space acts as a catalyst for disruptive experiences in virtual environments, a space for learning, innovation and knowledge production (Hollands, 2008; Carrillo, 2006).

Through virtual reality and geo-location apps making it possible to personalise all content, new flexible urban narratives are being offered adapted to the interests and concerns of different groups of the population (Koeck & Warnaby, 2015).

New smart approach is also evident in the multiplication of digital media formats and channels, segmented so that all urban stakeholders can consume and enjoy the city. The new technologies have brought speed, convenience and proximity, among other advantages, to everyday political communication techniques. A range of interactive communication possibilities has also opened up that would have been unthinkable with other tools and supports (Almansa & Castillo, 2014). Policy-makers, aware of the growing role of citizens both as generators of content about the urban area on social networks and in virtual spaces and, more traditionally, as co-creators of urban reality, are paying increasing attention to channels for communication with them. Rather than the normal pattern of a one-way, vertical monologue characteristic of communication between cities and their residents until the end of the 20th century, in which cities disseminated the information they considered relevant via the mass media and some of their owned media, such as newsletters or customer care lines (Graham, 2014), modern cities are choosing citizen relational, multi-directional communication models with active public involvement (Therkelsen, 2015; Campillo Alhama, 2012).

Together with city council websites, which have become the main portals for information and channels for interaction with citizens, it is increasingly common to see the active presence of city councils on the main social media such as Facebook, Twitter, YouTube and LinkedIn (Kolotouchkina & Moreno, 2016). Whatsapp and specific apps are also being used by City Councils to provide their citizens with a direct contact line and an immediate access to relevant information to carry out necessary procedures in cities (Iamsterdam.com, 2018).

Last, but not least, another relevant contribution of ICTs in the field of citizen engagement is linked to their unique role in enabling and facilitating social inclusion and access to information for citizens with disability. Although people with disability account for almost 15% of global population (World Health Organization (WHO), 2011), the public awareness, media representation and understanding about disability is yet superficial and stigmatized (Bush, Silk, Porter, & Howe, 2013). Negative attitudes, prejudices and stereotypes are commonly linked to the perception of people with disability, creating significant barriers to their full social engagement and access to key social experiences (WHO, 2011). Even though much work still needs to be done by urban policy-makers, meaningful ICTs led initiatives that facilitate social inclusivity and access to information resources for people with disability are emerging around the world in global cities.

Following these perspectives, the paper aims to:

1. Examine the structure and key features of the main digital communication platforms of smart cities.
2. Identify key tools put in place by smart cities, fostering citizen engagement and inclusivity.
3. Systematize the most innovative initiatives and best practices in the field of smart place making.

Building on the previous research on smart cities and citizen participation, this paper underlines the key role of citizen engagement and digital communication channels to building a truly inclusive, liveable and inspiring smart places.

BACKGROUND

Smart Place Making

The concept of smart city is usually approached form different schools of thought, polarised depending on the importance attributed to the power of ICTs to connect people, information and city elements with the main focus either on urban advanced efficiency and sustainability (Hall, 2000; Cretu, 2012; Nam & Pardo, 2011); their impact on urban efficiency in key service areas (Hajer & Dassen, 2014), or on fostering social capital, collective intelligence and spillover of knowledge (Caragliu et al., 2011; Komninos, 2011); the relationship between the city and its residents (Rheingold, 2002); and the generation of new narratives and virtual representations of the territory (Koeck & Warnaby, 2015; Blume & Langenbrick, 2004). Taking into consideration different existing approaches to the smart city concept, Giffinger et al. (2007) identify six main dimensions for smart development of the city, comprising both the technological and the human scale perspectives: smart economy, smart people, smart living, smart environment, smart mobility, and smart governance.

The technological perspective highlights the key role of ICTs and open data to address the quality of urban services, sustainability, safety and mobility (Hajer & Dassen, 2014; Picon, 2015). An extensive network of sensors and devices measuring, recording and connecting urban activities make the city become an innovative living lab (Sassen, 2011). Technological innovation and hyper-connectivity lead to important advances configuring a new context of connected, balanced, operational cities.

The effectiveness and operability in mobility and urban traffic is enhanced through the implementation of electronic pricing, additional rush-hour tolls, sensors in parking areas, the introduction of single cards for public transport and the intelligent programming of traffic flows at certain times of day. Advanced measurement and smart saving systems for the main commodities consumed, such as water, electricity and gas improve urban sustainability. Meanwhile, the gradual introduction of clean or renewable energy sources (solar, wind, geothermic and biomass), together with the integrated management and recycling of urban waste, make it possible to reduce the environmental impact of urban areas. The main tools developed by cities in the area of smart liveability include applications for cyber-security, video surveillance, the management of public alerts and urban geo-location. This same area also includes specific actions in smart public health management covering everything from the dealing with health warnings and developing remote care platforms for vulnerable and dependent people to telemedicine, digital rehabilitation, electronic medical records and advanced simulation in the training of medical and care staff (Telefonica, 2015).

On the other hand, a more holistic scenario of a smart city enhances its collective intelligence as a place of citizen´s inventiveness and creativeness (Capdevila & Zarlenga, 2015); a process of continuous learning and inspiration (Premalatha, Tauseef, Abbasi & Abbasi 2013); new mediated experience of social cooperation and mobilization (Rheingold, 2002) or knowledge spillover and knowledge-based urban development (Carrillo, 2006). At the same time, there is an increasing awareness of the impact of pervasive computing and mobile technologies on social and communication dynamics emerging within these smart urban spaces. The technology migration worldwide to both higher speed mobile and broadband networks together with an increased penetration of smartphones enable citizens to easily navigate through digital cityscapes and to become its active players and creators. The spread of social virtual networks, easily accessible through smartphones, replaces a direct physical interaction by creating new types of personal and collective behaviour as well as new formats and frameworks of interpersonal communication (Picon, 2015). The sharing of experiences, feedbacks, comments and insights about places through an extensive range of social networks is a common practice of a digital word of mouth nowadays.

Smart urban management concerning the population can be seen in the use of advanced methods and digital media for professional education and training. Open access to digital content, improving the skills of broad segments of the population so they can actively use ICTs and eliminating the digital gap affecting the most vulnerable or older segments of the population are among the challenges for city managers in this field. Furthermore, transparent, intelligent governance is supported by the creation of virtual offices providing public services, digital platforms for the integrated management of administrative procedures and transparency portals concerning urban planning, as well as providing access to the city's open data. Other significant actions include the implementation of urban wi-fi and the creation of specific apps, both by city authorities and citizens themselves, which are offered openly on municipal websites to facilitate participation and public connection with the main urban projects. Finally, the smart urban economy is advancing through collaborative economic and crowdsourcing projects; the creation of specialised technology start-up nurseries and clusters and the use of virtual maps and geolocation applications for the personalised management of cities' ranges of commercial, tourist and cultural attractions. Augmented reality, home automation applications, the Internet of Things, sensorisation, Open Data and cloud computing create a new context of smart enterprise in urban areas (Telefonica, 2015).

The World Health Organisation (2011) in its World Report on Disability recognizes that due to population ageing and increase in chronic health conditions, people with a health condition account for almost 15% of global population. The prevalence is higher among women, older people and lower income households. Cities play a critical role in fostering inclusivity and accessibility of their citizens with disability. Urban policies usually focus on two main fields of actions. Removing physical and technical barriers to free movement of people with disability and their physical access to urban services has become a key premise of building open and accessible cities for all. However, enabling access and providing adapted channels to government information, services, healthcare, employment and education opportunities, as well as to cultural content for citizens with disability, is especially important to address digital divide, social exclusion and economic hardship of citizens with disability (Mervyna, Simonb, & Allen, 2014; Gilbert, 2010).

The pace of urban technological innovation is uneven around the world. The large size of many modern cities, together with the inertia of urban development, often slow down the rapid adaptation of the best smart and sustainable urban development practices. However, an increasing numbers of cities are committed to getting the best from ICTs to make their urban spaces sustainable, balanced and inspiring places for their residents to live, work and play.

The case of new cities and urban districts entirely built under the concept of the smart management of all urban dynamics is another example of the importance of the holistic, integrating approach to the urban context (Kolotouchkina & Seisdedos, 2018).

PUBLIC PARTICIPATION: ACTIVE, CRITICAL AND CONNECTED CITIZENS

The importance of citizens in the development and growth of urban space is a recurring topic in academic research. Girardet (1999) highlights the impact of the active involvement of citizens in tackling urban sustainability problems. Peñalosa (2007) underlines the dimension of social equality as a key area for achieving the commitment of all groups of urban residents. Rogers & Gumuchdjian (1997) advocate the key role of citizens in channelling their physical, intellectual and creative energy to create a beautiful, just, green, compact, communicative and diverse city. Citizens are the heart and soul of a city. Through their customs and ways of acting and living, they define the unique character of an urban territory, its most authentic experiences and its identity. In the context of world globalisation that could potentially lead to greater cultural sameness (Pieterse, 1996), uniformity and the McDonaldization of society (Ritzer, 2019), cities concentrate a unique manifestation of the authenticity and singular nature of their residents (Sudjic, 2007).

From the communication perspective, citizens are the final recipients of the communication and information actions carried out by urban policy-makers. Citizens are the main target with which the government needs to connect to win their support in the key policy areas of economic, social and cultural affairs. At the same time, citizens make a significant contribution to urban identity and reputation through their personal content about their place of residence and work shared online. Joining digitally mediated networks has become a rewarding personal experience, as uploaded personal content is shared and recognized by others (Benkler, 2006). The impressive rates of active participation of people in social networks have created a new context of personal empowerment and participatory culture (Jenkins, 2006).

The unstoppable deployment of digital media and social networks makes citizens advocates and active narrators of the urban reality on account of their many personal and professional contacts (Kavaratzis, 2004). Statistics on the activity of the world population on social networks in 2019 show that almost six billion people have mobile connections, with 57% of these corresponding to smartphones, and about four billion people have internet access (GSMA, 2018; Internet Live Stats, 2019). Another outstanding feature is the number of users active on the different social networks. More than two billion people in the world are active users of Facebook, one billion use Instagram and 300 million regularly share content on Twitter and Pinterest (Internet Live Stats, 2019; Statista, 2018).

Bearing in mind that more than 50% of the world's population is currently concentrated in cities and that forecasts for 2050 point to this percentage reaching 70% (UN-Habitat, 2016), the role of urban residents as creators of content and data in the urban space is increasingly important. The context of hyper-connectivity encourages the replacement of the urban physical reality by virtual reality and mediated communication, creating new formats of interpersonal communication in virtual environments (Gumpert & Drucker, 2007; Picon, 2015). Sharing experiences, points of view and personal opinions on social media is increasingly common.

Participatory democracy is based on participation as a set of relational procedures and processes through which the agents enter into a symmetrical and reciprocal relationship of communication, cooperation and co-responsibility. The connection between the agents involved in social life from this participatory perspective, as a need and as a right, is what makes it possible to recover the transversal and relational meaning of participation. This requires innovations capable of translating participation into relational communication for joint action by citizens (Alguacil, 2006). Citizens cannot exercise their right to engage in public affairs and the decisions of the politicians who run the authorities and other public bodies if they do not have full, intelligible, transparent information about who those representatives are and whether they plan their actions to manage collective resources fairly and justly, effectively and efficiently. The plans they make guide what local authority officers and other employees must do (Molina, Corcoy-Rius, & Simelio-Sola, 2013).

The relevance of the participation and involvement of citizens in the different areas of urban management, as well as their triangular role in urban communication as recipients, creators and advocates is a relevant topic for smart place making research and practice.

METHOD

Based on a case study research method, the paper explores the most outstanding and innovative practices concerning the communication channels and formats implemented by two cities considered to be world leaders in smart urban place making. A case study method is frequently used in the assessment of the innovative projects and best practices in urban and communication research (Premalatha et al, 2013; Dinnie, 2013; Hajer & Dassen, 2014).

The sample was selected based on the review of the most relevant world rankings in the area of urban innovation and digitalisation: the Innovation Cities Index 2019; the European Digital Cities Index 2016-2017, and the Global Power Cities Index 2019.

The city of London is at the top positions in the latest editions of the three rankings consulted (MMF, 2020; Innovation Cities, 2020; European Digital City Index, 2016). The city of Madrid was included as the second element for analysis, considering the fact that the city had stood out ahead of other European cities in previous urban digital communication studies for its commitment to digitization of urban services, its innovative public participation initiatives and its high rates of activity and interaction on social media (Kolotouchkina and Moreno, 2016).

The content analysis carried out has focused specifically on exploring the best practices of the two city governments in the fields of smart place making through digital communication and public participation. The case study has been carried out using empirical analysis (Yin, 2009) to identify innovative strategies and best practices put in place by the governments of the two cities. Furthermore, a comparative analysis has been carried out to assess the performance of both cities. The unit for analysis in the research is provided by the websites of the city authorities, and particular attention has been paid to:

- the functionality and accessibility of digital communication channels;
- the key content and urban narratives displayed on those channels;
- the availability of specific tools fostering citizen participation in the running of the cities;
- the access to open data on urban dynamics.

Data and evidence for the analysis have been compiled by monitoring and analysing the content of the main digital communication platform of the municipal governments of London and Madrid during the last quarter of 2018. The presentation of each case follows the same structure in order to facilitate the comparative analysis of best practices in smart place making implemented by both cities: brief introduction; design, usability and tone of voice; accessibility tools; digital communication channels and public participation tools; multiplatform content; and open data.

It is important to acknowledge the limitations of this exploratory and interpretive case study methodology on the generalizability of the results. Thus, the conclusions should be considered as suggestive with these limitations and further research needs to be developed to assess the impact of ethical and socially relevant content in communication and advertising education.

SMART URBAN PLACE MAKING THROUGH DIGITAL COMMUNICATION AND CITIZEN ENGAGEMENT IN LONDON AND MADRID

London

The Greater London Authority website www.london.gov.uk is the city's main digital communication platform. The website combines information and contact details relating to the Mayor and the London Assembly, together with the presentation of different operational areas of the authority, links to the boroughs and their procedures, and the latest city news and events. As well as its purely informative function, the page is presented as a platform for interaction with the public, encouraging the participation of Londoners in a wide range of initiatives and events in the city.

Design, Usability and Tone of Voice

Visually, the structure of the content is clear, easily identifiable and well designed. Browsing is quick, simple and intuitive. The content is identified and can be accessed directly, without the need to download pdfs or use external links. The language used is friendly, direct and often colloquial, without technical terms or difficult words. The names of the different sections of the website use short sentences or calls to action: What we do; In my area; Get involved; About us; Talk London; Media Centre. The most commonly used grammatical forms in the content are the first person singular and plural and the second person singular.

Meanwhile, there is an important section about the city mayor, Sadiq Khan, which very clearly and directly presents his biography, salary and annual spending since his election, along with a detailed register of all gifts received during his term in office and the identification of their origin.

Accessibility Tools

Accessibility of the Greater London Authority website is provided through a series of specific tips for people with visual or hearing disabilities. The advice concerns the possibility of increasing the visibility and/or readability of the content, simplifying the use of the access computer keyboard and the option to hear all the content shown on the screen read aloud.

As the website is the main digital platform for the Greater London Authority, the contact section for the government of London is prominently featured. This section details all possible forms of direct communication with the mayor, the members of the city government and the media relations team. There is a list of the telephone numbers and personal e-mail addresses of all members of the government, specifying their names and corresponding areas of responsibility. As well as this list, through the website it is possible to sign up to the Mayor's personal mailing list to receive periodic updates and relevant content. The Mayor uses this direct, personalised communication channel, largely to announce his new projects and promote citizen participation in public debates and volunteering initiatives.

Digital Communication Channels and Public Participation Tools

A section of the municipal website focuses exclusively on public participation. The city offers different formulas for direct or indirect engagement in debates about government proposals, the distribution of the municipal budget and the implementation of important new initiatives. Outstanding among these is *Talk London*, an online community requiring registration, which offers Londoners the chance to take part in public debates on different urban projects and initiatives. In 2018, the community had more than 42,000 registered members. The projects approved with a public contribution via *Talk London* include: the city's Housing Strategy, the Smart City Plan and the public consultation on the environment (Mayor of London, London Assembly, 2018a). The same public area details the schedule for all meetings of the London Assembly and its various executive committees, which anyone can attend.

The meetings are broadcast live on the website and there is an archive of previous broadcasts.

Volunteering is another important part of the management of public participation in London. Through his personal programme, *Team London*, the mayor offers Londoners different volunteering options. Depending on the time they have available and their personal concerns, the programme allows residents to work with some of the more than 2,000 non-governmental organisations registered in London or to become city ambassadors to support tourists and visitors at big sporting and cultural events. It offers executives with broad business experience the chance to contribute to enterprise training for students at schools in the city (Mayor of London, London Assembly, 2018b).

Multiplatform Content

The interactive nature of the website is shown in the use of specific tags associated with the main events in the British capital. The use of tags makes it easier to relate content on different digital platforms and expands their impact and importance on the internet.

The main hashtag on the website is #LondonisOpen, from the communication campaign launched by Sadiq Khan after the referendum on whether the United Kingdom should remain in the European Union (Mayor of London, London Assembly, 2018c). The campaign reaffirms the enterprising, international and creative spirit of a city that is home to a large number of immigrants. A wide selection of audiovisual testimonials from different kinds of residents, including artists, small business owners, elite sportspeople, students and immigrant families, shows different points of view on the city's vibrant character, its multicultural fusion and its tolerance. Other important hashtags are #LondonUnited, representing the municipal initiative to support the victims of the terrorist attacks that recently took place in the city, and #BehindEveryGreatCity, devoted to the centenary of the movement demanding votes for women. Meanwhile, #MyLondyn identifies all the content on the commemoration of the city's Polish community.

The Greater London Authority website operates as a link to the profiles of the Mayor and his governing team on the main social media: Facebook, Twitter, Instagram and YouTube. These profiles are used to share the most important events, news and content concerning the city in line with the specific nature of each medium. There is also a Greater London Authority blog used to offer broader content in terms of subject matter and importance for different kinds of citizen. This content includes events commemorating the victims of terrorist attacks or serious accidents occurring in London; interviews with citizens and immigrants; and a range of advice, such as how to behave with an abusive landlord, how to protect yourself in a heat wave and how to observe the full moon.

Open Data

Associated with the government of London website is the city's open data portal with all kinds of figures, statistical indicators, sector analysis and themed reports (Mayor of London, London Assembly, 2018d). The information is classified into 17 categories, ranging from very general ones (population, transparency, education, transport, housing, culture and the environment) to more specific headings like crime and public safety, young people, London 2012, income, poverty and welfare, among others. The content is segmented depending on whether it relates to the city in general or to the various boroughs. There are more than 32 different categories of geographical segmentation of the information. More than 60 public institutions, businesses and organisations appear as providers of all the data, reports and analysis. Access to it is easy and intuitive and the content can be downloaded in 12 different storage formats.

Madrid

The Madrid City Council website www.madrid.es is a municipal digital communication platform with a wide variety of content and functions. Together with the information about the city and its government, the portal presents a wide range of procedures, resources and segmented content depending on the user's profile and their areas of interest.

Design, Usability and Tone of Voice

Visually, the structure of the Madrid City Council website is clear and simple, and the different levels of municipal content depending on user's specific needs or profile can easily be identified. The main feature of the principal browsing bar is the section on municipal procedures, followed by the news section, the presentation of the city council and the districts of the city, and the contact information. The procedures section is a direct link to the city council's digital office which allows a wide range of online procedures. There, citizens can make appointments to pay bills and taxes, obtain certificates, give notification of incidents and request municipal services.

Tabs with pictograms and their corresponding descriptors provide immediate information about the themed content relating to the nine municipal management areas, including economic activity and finance; emergencies and safety; culture; sport and leisure; the environment and mobility and transport). The page also provides direct access to the main news about the city, the events calendar, online procedures, public communication channels and the open data portal.

The language used on the website is clear and simple. The names of all the sections are highly descriptive, without technical terms or words that are difficult to understand. The content is largely written in the second person singular or a neutral, descriptive style.

The section about the mayor is summarised in a personal greeting, with a brief message about her vision of the city.

Accessibility Tools

The Madrid City Council website has been designed and programmed in accordance with the accessibility standards of the W3C, the world authority for accessible digital content. These say that websites should be perceivable, operable, understandable and robust (W3C, 2018). Browsing offers the user the chance to choose the size of the text to make it easier to read. People with visual disabilities can listen to the content using the voice synthesiser tool. There are also browsing key shortcuts for the main content blocks. The website is structured in cascade to make it easy to read and access from different browsers.

One of the city council's outstanding initiatives in the area of accessibility is its Clear Communication project, reflecting the fundamental premise that citizens are entitled to understand communication from public institutions without obstacles. Its main aim is to promote the use of clear, direct, simply, comprehensible language on all channels and media for communicating with the public. Meanwhile, wherever possible, an attempt is made to include easy reading, allowing the content to be understood by people with learning difficulties (Madrid City Council website, 2018a).

Digital Communication Channels and Public Participation Tools

Madrid City Council's leading initiative in terms of public participation is the open government portal *Decide Madrid*. The platform was set up to implement a direct, personalised, participatory communication channel to present citizen initiatives, participatory budgeting proposals, voting on various initiatives and the evaluation of their subsequent implementation. Registration on the website to take part in debates and create new proposals is available to all citizens, but statements of personal support and final voting on the initiatives presented are restricted to all those registered as residents of the Spanish capital. The mechanics of public participation are very simple and open to all kinds of suggestions and proposals. However, moving a vote on any proposal requires the support of 1% of people aged over 16 registered as residents in Madrid. If the proposal is approved in the final public vote, the city council plans its implementation. Since 2017, there have been public votes for the *Sustainable Madrid* project and the introduction of a single public transport card for the city. Meanwhile, the city council has carried out public consultation on its big urban planning projects using this channel, such as the remodelling of Plaza de España, Calle Gran Vía and various squares in the city (Decide Madrid, 2018).

Meanwhile, every year between January and February the citizens registered as residents in the capital are asked to present their proposals on the distribution of 100,000 million euros from participatory budgeting assigned to finance projects of interest for certain districts or for Madrid as a whole. Each project presented goes successively through the stages of public exhibition to achieve support, evaluation by council officers, final voting on the selected projects and the final presentation of the projects receiving most votes.

Another initiative on the *Decide Madrid* portal focuses on debates opened up among citizens about subjects that concern and interest them. Each debate allows comments and votes from participants.

At the same time, the *Participatory Process* section offers the people of Madrid the chance to present their comments, suggestions and observations about the various municipal strategies, rules or projects. The website clearly identifies the period for commenting and includes all the comments made so far. Multiplatform Content

Línea Madrid is Madrid City Council's main public services channel. It includes all means of communication with the public via various digital and face-to-face media: the 010 public services telephone line; network of *Línea Madrid* physical offices, the city council's website and its Twitter profile @Lineamadrid. *Línea Madrid* provides information about the city council and the city of Madrid and makes the various procedures the public have to carry out with the public authorities easier. These public procedures mainly involve education and culture, mobility, social services, housing, the register of residents, public participation and the environment. Throughout 2017, *Línea Madrid* provided more than 10 million public services via its digital and physical channels and has more than 19 million visits to its website (Madrid City Council website, 2018b).

Another important feature of Madrid City Council's multiplatform content is its *Avisos Madrid* app for mobile devices. This application was created to provide information about incidents in the capital concerning street furniture, cleaning, vehicles or street lighting, for example, together with follow-up.

The app makes it possible to take a photograph of the incident, geolocate it and describe it. Once the incident has been recorded, it is automatically registered in the city council's Alert Management System. The app has been designed as a social community based on these alerts, and also offers the option to make requests for the installation of new urban street furniture, cleaning and the collection of old furniture, as well as to access all the city council's communication channels and the 010 line, the public services profile on Twitter@Lineamadrid_and the city council's website (Madrid City Council website, 2018c)._

Madrid City Council has a digital publication called *Diario Madrid,* which offers a summary of the main events and news from the city, together with the city council press releases, the presentation of useful advice, interviews and a wide variety of municipal initiatives (News from Madrid City Council, 2018). Furthermore, city council has profiles on four social networks: Twitter, Facebook, Instagram and YouTube, although these work in different ways. The @Lineamadrid Twitter profile is used as a communication and public services channel, offering the option to carry out certain procedures, such as requesting access to residents' priority areas in Madrid and the collection of old furniture. Facebook, Instagram and YouTube profiles act as news channels and amplifiers for the main news and events in the capital.

Open Data

The data relating to Madrid City Council's administration is included on two different websites: the Transparency portal and the Open Data portal. The Transparency portal offers a wide variety of content with the ultimate aim of providing clarity and absolute transparency in the management of all areas of the city's government. The main sections of the website include the content relating to the distribution of the municipal budget, legal regulations, the human resources policy, citizen relations, environmental management and urban planning. The data provided includes the salaries of all members of the governing team, the lobby register; policy and regulations; a breakdown of the budget; and an inventory or list of collaboration agreements signed by the city council with public or private bodies (Transparency portal of Madrid City Council, 2018).

Madrid City Council also has a website specialising in open data. The aim of this portal is to make it easier to access public data in reusable formats to expand and enrich municipal content and generate new applications, services and business projects. The portal offers 343 sets of data relating to different areas of the city, with 2,599 downloadable resources and 4,460 reuses (Open Data portal of Madrid City Council, 2018).

CONCLUSION

The study has made it possible to identify a wide variety of digital communication channels linked to the municipal websites. In both London and Madrid, the municipal authorities' websites combine multiple roles: image, information, procedures, links to other relevant content and platforms for interacting with the public. Meanwhile, there is a combination of multimedia content on both municipal websites. The main digital communication channels of both city authorities are outstandingly multifaceted, multifunctional and multimedia.

At the same time, the tone and style of the content of the municipal websites is also important. In both cases, the use of simple, easy language has been detected, avoiding technical terms, abbreviations and words that are difficult to understand. Along these lines, Madrid City Council places special emphasis on its *Clear Communication* programme to avoid any confusion in the municipal content and make access easier for people with learning difficulties.

Most of the content is written in the first and second person singular and plural, directly involving users of the channel in a friendly, open conversation. In the case of the Greater London Authority, most sections of the municipal website are written in a friendly, colloquial style.

Another outstanding feature of the municipal websites analysed is their commitment to accessibility and inclusivity of people with disability. In both cases, specific tools and assistive technologies have been identified to make the content easier to read, display, listen to and understand.

Both city authorities have set up channels for the public to participate in the running of their respective cities. The *Talk London* and the *Decide Madrid* portals offer a wide variety of options for citizens to participate in the running of the city, monitor the municipal government, publicise their opinions about urban projects, present new improvement proposals and open up a public debate on the most important issues. The participatory budgeting project implemented by Madrid City Council in the context of its *Decide Madrid* portal is the most important example of direct public involvement in urban management.

A comparative analysis of the best practices in smart place making through digital communication and public participation tools developed by the municipal authorities in London and Madrid shows a conscious attitude of the governments of both cities to encourage permanent dialogue with their citizens, fostering their active public involvement. Furthermore, the analysis allows identifying some specific drivers of citizen participation common to both cities. On one hand, these drivers vary significantly on their level of engagement, from active to passive participation, allowing citizens to choose the role they prefer to take, if any, in a particular urban experience. The experience value provided by these drivers ranges from access to a specific knowledge and expertise to community empowerment, information or pure entertainment, social connections and easy access to city services.

The commitment of both city authorities to an active social media presence is outstanding, and they operate an integrated multichannel communication strategy. Municipal government profiles have been found on the main social media in which content is updated in the specific format of each network. Madrid City Council's

Table 1. Comparative analysis of digital communication and public participation channels of the municipal authorities in London and Madrid

Analysis criteria	London	Madrid
Design, usability and tone of voice	Clear, simple structure Easy browsing: access different content levels in just a few clicks Direct, friendly language	
	Content written in the first person singular and plural and the second person singular	Content written in the second person singular or in a neutral style Clear Communication project
Accessibility tools	Option to increase the visibility and/or readability of the content Option to simplify the use of the access computer keyboard Option to listen to the website content being read aloud	
Digital communication channels and public participation tools	Talk London portal Team London project Mayor's personal mailing list Twitter and Facebook profiles of the Mayor and his team	Decide Madrid portal Avisos Madrid app City council's Twitter profile
Multiplatform content Social networks	#LondonisOpen #LondonUnited #Behindeverygreatcity #MyLondyn City Hall blog Twitter Facebook Instagram YouTube	Línea Madrid Avisos Madrid app *Diario Madrid* Twitter Facebook Instagram YouTube
Availability of open data	Open data portal	Transparency portal Open data portal

Source: own creation

Twitter profile is used not only as a communication channel but also to manage temporary parking permits and resolve queries from the public.

Finally, the commitment to transparency in urban management is important in both cases analysed via open data portals concerning the cities and their management. There is also detailed communication on municipal websites of the budgets approved and their implementation is monitored.

In summary, the analysis of the London and Madrid municipal websites has made it possible to identify a series of relevant practices in digital communication and citizen participation, which are important in achieving active public engagement in managing and creating the smart urban space. The results of the comparative analysis show quite a similar approach from two cities both at strategic level and in specific initiatives carried out. The city authorities are committed to transparency, permanent dialogue, accessibility and public inclusiveness in their administration. The general approach and common experiences identified in the cases of London and Madrid point to the consolidation of a new context for communication and urban management, with citizens becoming key agents of the smart urban space.

REFERENCES

W3C. (2018). *Introduction to Web Accessibility*. Retrieved from https://www.w3.org/WAI/fundamentals/accessibility-intro/

Albino, V., Berardi, U., & Dangelico, R. M. (2015). Smart Cities: Definitions, Dimensions, Performance, and Initiatives. *Journal of Urban Technology*, *22*(1), 3–21. doi:10.1080/10630732.2014.942092

Alguacil Gómez, J. (2006). Los desafíos del nuevo poder local: ¿hacia una estrategia relacional y participativa en el gobierno de la ciudad? In J. Alguacil Gómez (Ed.), *Poder local y participación democrática*. Barcelona: El Viejo Topo.

Almansa, A., & Castillo–Espacia, A. (2014). Comunicación Institucional en España. Estudio del uso que los diputados españoles hacen de las TIC en sus relaciones con la ciudadanía. *Chasqui*, *26*(126). doi:10.16921/chasqui.v0i126.250

Benkler, Y. (2006). *The wealth of networks: how social production transforms markets and freedom*. New Haven: Yale University Press.

Blume, T., & Langenbrinck, G. (2004). *Dot.city. Relational Urbanism and New Media*. Berlin: Jovis.

Bush, A., Silk, M., Porter, J., & Howe, P. D. (2013). Disability [sport] and discourse: Stories within the Paralympic legacy. *Reflective Practice*, *14*(5), 632–647. doi:10.1080/14623943.2013.835721

Campillo Alhama, C. (2012). Investigación en comunicación municipal: Estudios y aportaciones académicas. *Vivat Academia*, *117E*(117E), 301–323. doi:10.15178/va.2011.117E.1035-1048

Capdevila, I., & Zarlenga, M. I. (2015). Smart city or smart citizens? The Barcelona case. *Journal of Strategy and Management*, *8*(3), 266–282. doi:10.1108/JSMA-03-2015-0030

Caragliu, A., Del Bo, C., & Nijkamp, P. (2011). Smart Cities in Europe. *Journal of Urban Technology*, *18*(2), 65–82. doi:10.1080/10630732.2011.601117

Carrillo, F. J. (2006). *Knowledge Cities: Approaches, Experiences and Perspectives*. Elsevier Butterworth Heinemann. doi:10.4324/9780080460628

Castells, M. (1996). *The Rise of the Network Society. The Information Age: Economy, Society and Culture* (Vol. 1). Chichester, UK: John Wiley & Sons.

Cretu, G. I. (2012). Smart Cities Design Using Event-driven Pradigm and Semantic Web. *Informações Econômicas, 16*(4), 57–67.

Decide Madrid. (2018). *Presupuestos participativos*. Retrieved from https://decide.madrid.es/presupuestos

Dinnie, K. (2011). *City branding. Theory and cases*. London: Palgrave MacMillan. doi:10.1057/9780230294790

European Digital City Index. (2016). *European Digital City Index 2016*. Retrieved from https://digitalcityindex.eu/

Giffinger, R., Fertner, C., Kramar, H., Kalasek, R., Pichler-Milanović, N., & Meijers, E. (2007). *Smart Cities: Ranking of European Medium-sized Cities*. Academic Press.

Gilbert, M. (2010). Theorizing digital and urban inequalities. Critical geographies of 'race', gender and technological capital. *Information Communication and Society, 13*(7), 13, 1000–1018. doi:10.1080/1369118X.2010.499954

Girardet, H. (1999). Creating Sustainable Cities, Schumacher Briefings n° 2 (6th ed.). Green Books Ltd.

Govers, R. (2015). Rethinking Virtual and Online Place Branding. In *Rethinking Place Branding*. Springer International Publishing. doi:10.1007/978-3-319-12424-7_6

Graham, M. (2014). Government Communication in the Digital Age: Social Media's Effect on Local Government Public Relations. *Public Relations Inquiry, 3*(3), 361–376. doi:10.1177/2046147X14545371

GSMA. (2018). *The Mobile Economy 2018*. Retrieved from https://www.gsma.com/mobileeconomy/wp-content/uploads/2018/05/The-Mobile-Economy-2018.pdf

Hajer, M. A. & Dassen, T. (2014). *Smart About Cities: Visualising the Challenge for 21st-Century Urbanism*. Rotterdam: nnai010publishers/PBL publishers.

Hall, R. E. (2000). The vision of Smart City. *Proceedings of the 2nd Inteernational Life extension Technology Workshop*.

Hollands, R. G. (2008). Will the real smart city please stand up? *City, 12*(3), 303–320. doi:10.1080/13604810802479126

Iamsterdam. (2018). *The City of Amsterdam and Whatsapp*. Retrieved from https://www.iamsterdam.com/en/living/latest-news/whatsapp-the-city

Innovation Cities. (2020). *Innovation Cities Index 2019.* Retrieved from https://www.innovation-cities.com/index-2019-global-city-rankings/18842/

Internet Live Stats. (2019). Retrieved from https://www.internetlivestats.com/

Jenkins, H. (2006). *Convergence Culture: Where Old and New Media Collide.* New York: New York University Press.

Kavaratzis, M. (2004). From city marketing to city branding: Towards a theoretical framework for developing city brands. *Place Branding and Public Diplomacy, 1*(1), 58–73. doi:10.1057/palgrave.pb.5990005

Koeck, R., & Warnaby, G. (2015). Digital Chorographies: Conceptualising Experiential Representation and Marketing of Urban/Architectural Geographies, *Architectural. Research Quarterly, 19*(2), 183–192.

Kolotouchkina, O., & Moreno, P. (2016). Comunicación digital urbana: retos y perspectivas. Análisis comparativo de seis capitales europeas. In M. Linares Herrera, J. Diaz Cuesta, & M.E. Del Valle Mejías (Eds.), Innovación universitaria: digitalización 2.0 y excelencia en contenidos. Madrid: McGraw Hill Education.

Kolotouchkina, O., & Seisdedos, G. (2018). Place Branding Strategies in the Context of New Smart Cities: Songdo IBD, Masdar and Skolkovo. *Place Branding and Public Diplomacy, 14*(2), 115–124. doi:10.105741254-017-0078-2

Komninos, N. (2011). Intelligent Cities: Variable Geometries of Spatial Intelligence. *Intelligent Buildings International, 3*(3), 172–188. doi:10.1080/17508975.2011.579339

Madrid City Council open data portal. (2018). *Datos abiertos.* Retrieved from https://datos.madrid.es/portal/site/egob/

Madrid City Council transparency portal. (2018). Retrieved from https://transparencia.madrid.es/portal/site/transparencia

Madrid City Council website. (2018a). *¿Qué es la Comunicación clara?* Retrieved from https://www.madrid.es/portales/munimadrid/es/Inicio/Actualidad/Comunicacion-Clara/?vgnextfmt=default&vgnextoid=a01f1905bacde510VgnVCM1000001d4a900aRCRD&vgnextchannel=59af566813946010VgnVCM100000dc0ca8c0RCRD&idCapitulo=10506829

Madrid City Council website. (2018b). *¿Qué es Línea Madrid?* Retrieved from: https://www.madrid.es/portales/munimadrid/es/Inicio/El-Ayuntamiento/Atencion-a-la-ciudadania/Que-es-Linea-Madrid-/Que-es-Linea-Madrid-/?vgnextfmt=default&vgnextoid=a7d8ed76696f5310VgnVCM1000000b205a0aRCRD&vgnextchannel=95262bb1676f5310VgnVCM1000000b205a0aRCRD

Madrid City Council website. (2018c). *Aplicación móvil-Avisos Madrid*. Retrieved from: https://www.madrid.es/portales/munimadrid/es/Inicio/El-Ayuntamiento/Atencion-a-la-ciudadania/Aplicacion-movil-Avisos-Madrid/Aplicacion-movil-Avisos-Madrid/?vgnextfmt=default&vgnextoid=fcccd2ee88ff1610VgnVCM1000001d4a900aRCRD&vgnextchannel=517cd2ee88ff1610VgnVCM1000001d4a900aRCRD

Mayor of London. London Assembly. (2018a). *Talk London*. Retrieved from https://www.london.gov.uk/talk-london/

Mayor of London. London Assembly. (2018b). *About Team London*. Retrieved from https://www.london.gov.uk/what-we-do/volunteering/about-team-london-0

Mayor of London. London Assembly. (2018c). *Our #LondonisOpen campaign*. Retrieved from https://www.london.gov.uk/about-us/mayor-london/londonisopen

Mayor of London. London Assembly. (2018d). *London data store*. Retrieved from https://data.london.gov.uk/dataset

Mervyna, K., Simonb, A., & Allen, D. K. (2018). Digital inclusion and social inclusion: A tale of two cities. *Information Communication and Society*, *17*(9), 1086–1104. doi:10.1080/1369118X.2013.877952

MMF. (2020). *Global Power City Index*. Retrieved from http://www.mori-m-foundation.or.jp/english/ius2/gpci2/

Molina Rodríguez-Navas, P., Corcoy-Rius, M., & Simelio-Solà, N. (2015). Evaluación de transparencia y calidad de la información de las entidades no lucrativas. In P.Molina Rodríguez-Navas (Ed.), Transparencia de la comunicación pública local: el Mapa Infopaticipa. Sociedad Latina de Comunicación Social.

Nam, T., & Pardo, T. A. (2011). Conceptualizing Smart City with Dimensions of Technology, People, and Institutions. *Proceedings 12th Conference on Digital Government Research*. 10.1145/2037556.2037602

News from Madrid City Council. (2018). *Diario Madrid*. Retrieved from https://diario.madrid.es/

Peñalosa, E. (2007). Politics, Power, Cities. In *The Endless City, The Urban Age Project by the London School of Economics and Deutsche Bank's Alfred Herrhausen Society* (pp. 307–319). London: Phaidon Press Ltd.

Picon, A. (2015). Smart Cities. A Spatialized Intelligence. New York: Palgrave Macmillan. doi:10.1002/9781119075615

Pieterse, J. (1996). Globalisation and culture. Three paradigms. *Economic and Political Weekly*, *31*(23), 1389–1393.

Premalatha, M., Tauseef, S. M., Abbasi, T., & Abbasi, S. A. (2013). The promise and the performance of the world's ŏrst two zero carbon eco-cities. *Renewable & Sustainable Energy Reviews*, *25*, 660–669. doi:10.1016/j.rser.2013.05.011

Rheingold, H. (2002). *Smart Mobs: the Next Social Revolution*. Cambridge, MA: Perseus.

Ritzer, G. (2019). *The McDonaldization of Society: Into the digital age*. Thousand Oaks, CA: SAGE.

Rogers, R., & Gumuchdjian, P. (1997). *Cities for a small planet*. London: Faber and Faber.

Sassen, S. (2011). Talking back to your intelligent city. *McKinsey & Company*. Retrieved from http://voices.mckinseyonsociety.com/talking-back-toyour-intelligent-city/

Statista. (2018). *Number of Monthly Active Instagram Users from January 2013 to June 2018 (in millions)*. Retrieved from https://www.statista.com/statistics/253577/number-of-monthly-active-instagram-users/

Sudjic, D. (2007). Theory, policy and practice. In R. Burdett & D. Sudjic (Eds.), *The Endless City, The Urban Age Project by the London School of Economics and Deutsche Bank's Alfred Herrhausen Society* (pp. 32–50). London: Phaidon Press Ltd.

Telefonica. (2015). *Smart cities. La Transformación digital urbana*. Retrieved from https://iot.telefonica.com/libroblanco-smart-cities/media/libro-blanco-smart-cities-esp-2015.pdf

Therkselsen, A. (2015). Rethinking Place Brand Communication: From Product-Oriented Monologue to Consumer-Engaging Dialogue. In *Rethinking Place Branding* (pp. 159–173). Springer.

UN-Habitat. (2016). *World Cities Report*. Nairobi: UN-Habitat.

WHO. (2011). *World report on disability*. Retrieved from https://www.who.int/disabilities/world_report/2011/report.pdf

Yin, R. K. (2009). *Case Study Research Design and Methods*. Los Angeles: SAGE.

Related Readings

To continue IGI Global's long-standing tradition of advancing innovation through emerging research, please find below a compiled list of recommended IGI Global book chapters and journal articles in the areas of green cities, environmental management, and sustainable urban development. These related readings will provide additional information and guidance to further enrich your knowledge and assist you with your own research.

Abu-Shanab, E. A. (2017). E-government Contribution to Better Performance by Public Sector. *International Journal of Electronic Government Research*, *13*(2), 81–96. doi:10.4018/IJEGR.2017040105

Abu-Shanab, E. A., & Osmani, M. (2019). E-Government as a Tool for Improving Entrepreneurship. *International Journal of Electronic Government Research*, *15*(1), 36–46. doi:10.4018/IJEGR.2019010103

Adria, M., Messinger, P. R., Andrews, E. A., & Ehresman, C. (2020). Participedia as a Ground for Dialogue. In M. Adria (Ed.), *Using New Media for Citizen Engagement and Participation* (pp. 219–239). Hershey, PA: IGI Global. doi:10.4018/978-1-7998-1828-1.ch012

Ahrari, S., Zaremohzzabieh, Z., & Abu Samah, B. (2017). Influence of Social Networking Sites on Civic Participation in Higher Education Context. In R. Luppicini & R. Baarda (Eds.), *Digital Media Integration for Participatory Democracy* (pp. 66–86). Hershey, PA: IGI Global. doi:10.4018/978-1-5225-2463-2.ch004

Ahuja, D. (2018). Child Labour: Psychological and Behavioral Challenges. In S. Chhabra (Ed.), *Handbook of Research on Civic Engagement and Social Change in Contemporary Society* (pp. 238–252). Hershey, PA: IGI Global. doi:10.4018/978-1-5225-4197-4.ch014

Al-Emadi, A., & Anouze, A. L. (2018). Grounded Theory Analysis of Successful Implementation of E-Government Projects: Exploring Perceptions of E-Government Authorities. *International Journal of Electronic Government Research, 14*(1), 23–52. doi:10.4018/IJEGR.2018010102

Al-Jamal, M., & Abu-Shanab, E. (2018). Open Government: The Line between Privacy and Transparency. *International Journal of Public Administration in the Digital Age, 5*(2), 64–75. doi:10.4018/IJPADA.2018040106

Al-Ma'aitah, M. (2019). Drivers of E-Government Citizen Satisfaction and Adoption: The Case of Jordan. *International Journal of E-Business Research, 15*(4), 40–55. doi:10.4018/IJEBR.2019100103

Al-Yafi, K. (2019). A Quantitative Evaluation of Costs, Opportunities, Benefits, and Risks Accompanying the Use of E-Government Services in Qatar. In A. Molnar (Ed.), *Strategic Management and Innovative Applications of E-Government* (pp. 200–228). Hershey, PA: IGI Global. doi:10.4018/978-1-5225-6204-7.ch009

Alcaide-Muñoz, C., Alcaide-Muñoz, L., & Alcaraz-Quiles, F. J. (2018). Social Media and E-Participation Research: Trends, Accomplishments, Gaps, and Opportunities for Future Research. In L. Alcaide-Muñoz & F. Alcaraz-Quiles (Eds.), *Optimizing E-Participation Initiatives Through Social Media* (pp. 1–27). Hershey, PA: IGI Global. doi:10.4018/978-1-5225-5326-7.ch001

Alguliyev, R. M., & Aliguliyev, R. M., & Niftaliyeva (Iskandarli), G. Y. (2019). A Method for Social Network Extraction From E-Government. *International Journal of Information Systems in the Service Sector, 11*(3), 37–55. doi:10.4018/IJISSS.2019070103

Alguliyev, R. M., Aliguliyev, R. M., & Niftaliyeva, G. Y. (2018). Filtration of Terrorism-Related Texts in the E-Government Environment. *International Journal of Cyber Warfare & Terrorism, 8*(4), 35–48. doi:10.4018/IJCWT.2018100103

Alguliyev, R. M., & Yusifov, F. F. (2018). The Role and Impact of Social Media in E-Government. In L. Alcaide-Muñoz & F. Alcaraz-Quiles (Eds.), *Optimizing E-Participation Initiatives Through Social Media* (pp. 28–53). Hershey, PA: IGI Global. doi:10.4018/978-1-5225-5326-7.ch002

Alsaç, U. (2017). EKAP: Turkey's Centralized E-Procurement System. In R. Shakya (Ed.), *Digital Governance and E-Government Principles Applied to Public Procurement* (pp. 126–150). Hershey, PA: IGI Global. doi:10.4018/978-1-5225-2203-4.ch006

Alsaif, F., & Vale, B. (2018). The Use of Social Media in Facilitating Participatory Design: A Case Study of Classroom Design. In L. Alcaide-Muñoz & F. Alcaraz-Quiles (Eds.), *Optimizing E-Participation Initiatives Through Social Media* (pp. 209–235). Hershey, PA: IGI Global. doi:10.4018/978-1-5225-5326-7.ch009

Aluko, O. I. (2019). Trust and Reputation in Digital Environments: A Judicial Inkling on E-Governance and M-Governance. In R. Abassi & A. Ben Chehida Douss (Eds.), *Security Frameworks in Contemporary Electronic Government* (pp. 191–206). Hershey, PA: IGI Global. doi:10.4018/978-1-5225-5984-9.ch009

Aluko, O. I., & Aderinola, G. T. (2019). E-Governance and Corruption Impasse in Nigeria: A Developmental Expedition Synopsis. In R. Abassi & A. Ben Chehida Douss (Eds.), *Security Frameworks in Contemporary Electronic Government* (pp. 129–149). Hershey, PA: IGI Global. doi:10.4018/978-1-5225-5984-9.ch006

Anthony, S. M., & Zhang, W. (2017). Alternative Tweeting: A Comparison of Frames in Twitter's Political Discourse and Mainstream Newspaper Coverage of the Singapore General Election of 2011. In M. Adria & Y. Mao (Eds.), *Handbook of Research on Citizen Engagement and Public Participation in the Era of New Media* (pp. 324–343). Hershey, PA: IGI Global. doi:10.4018/978-1-5225-1081-9.ch018

Assay, B. E. (2019). Cyber Crime and Challenges of Securing Nigeria's Cyber-Space Against Criminal Attacks. In R. Abassi & A. Ben Chehida Douss (Eds.), *Security Frameworks in Contemporary Electronic Government* (pp. 150–172). Hershey, PA: IGI Global. doi:10.4018/978-1-5225-5984-9.ch007

Augustine, T. A., & Redman, D. P. (2019). Using Critical Literacy Skills to Support Civic Discourse. In A. Cartwright & E. Reeves (Eds.), *Critical Literacy Initiatives for Civic Engagement* (pp. 1–28). Hershey, PA: IGI Global. doi:10.4018/978-1-5225-8082-9.ch001

Baarda, R. (2017). Digital Democracy in Authoritarian Russia: Opportunity for Participation, or Site of Kremlin Control? In R. Luppicini & R. Baarda (Eds.), *Digital Media Integration for Participatory Democracy* (pp. 87–100). Hershey, PA: IGI Global. doi:10.4018/978-1-5225-2463-2.ch005

Babaoglu, C., & Akman, E. (2018). Participation With Social Media: The Case of Turkish Metropolitan Municipalities in Facebook. In L. Alcaide-Muñoz & F. Alcaraz-Quiles (Eds.), *Optimizing E-Participation Initiatives Through Social Media* (pp. 77–102). Hershey, PA: IGI Global. doi:10.4018/978-1-5225-5326-7.ch004

Bargh, M. S., Choenni, S., & Meijer, R. F. (2017). Integrating Semi-Open Data in a Criminal Judicial Setting. In C. Jiménez-Gómez & M. Gascó-Hernández (Eds.), *Achieving Open Justice through Citizen Participation and Transparency* (pp. 137–156). Hershey, PA: IGI Global. doi:10.4018/978-1-5225-0717-8.ch007

Barral-Viñals, I. (2017). Consumer "Access to Justice" in EU in Low-Value Cross-Border Disputes and the Role of Online Dispute Resolution. In C. Jiménez-Gómez, & M. Gascó-Hernández (Eds.), Achieving Open Justice through Citizen Participation and Transparency (pp. 191-208). Hershey, PA: IGI Global. doi:10.4018/978-1-5225-0717-8.ch010

Bawack, R. E., Kamdjoug, J. R., Wamba, S. F., & Noutsa, A. F. (2018). E-Participation in Developing Countries: The Case of the National Social Insurance Fund in Cameroon. In L. Alcaide-Muñoz & F. Alcaraz-Quiles (Eds.), *Optimizing E-Participation Initiatives Through Social Media* (pp. 126–154). Hershey, PA: IGI Global. doi:10.4018/978-1-5225-5326-7.ch006

Bessant, J. (2017). Digital Humour, Gag Laws, and the Liberal Security State. In R. Luppicini & R. Baarda (Eds.), *Digital Media Integration for Participatory Democracy* (pp. 204–221). Hershey, PA: IGI Global. doi:10.4018/978-1-5225-2463-2.ch010

Briziarelli, M., & Flores, J. (2018). Mediation Is the Message: Social Media Ventures in Informational Capitalism. In S. Chhabra (Ed.), *Handbook of Research on Civic Engagement and Social Change in Contemporary Society* (pp. 311–327). Hershey, PA: IGI Global. doi:10.4018/978-1-5225-4197-4.ch018

Cano, J., & Hernández, R. (2017). Managing Software Architecture in Domains of Security-Critical Systems: Multifaceted Collaborative eGovernment Projects. In S. Zoughbi (Ed.), *Securing Government Information and Data in Developing Countries* (pp. 1–26). Hershey, PA: IGI Global. doi:10.4018/978-1-5225-1703-0.ch001

Cano, J., Pomed, L., Jiménez-Gómez, C. E., & Hernández, R. (2017). Open Judiciary in High Courts: Securing a Networked Constitution, Challenges of E-Justice, Transparency, and Citizen Participation. In C. Jiménez-Gómez & M. Gascó-Hernández (Eds.), *Achieving Open Justice through Citizen Participation and Transparency* (pp. 36–54). Hershey, PA: IGI Global. doi:10.4018/978-1-5225-0717-8.ch003

Chania, A., & Demetrashvili, K. (2017). Public Procurement Reform in Georgia: The Way from Paper-Based Procurement to an E-Procurement System. In R. Shakya (Ed.), *Digital Governance and E-Government Principles Applied to Public Procurement* (pp. 151–169). Hershey, PA: IGI Global. doi:10.4018/978-1-5225-2203-4.ch007

Chapman, C. (2018). Student Acceptance of a Civic Engagement Graduation Requirement in an Urban Community College. In S. Chhabra (Ed.), *Handbook of Research on Civic Engagement and Social Change in Contemporary Society* (pp. 40–62). Hershey, PA: IGI Global. doi:10.4018/978-1-5225-4197-4.ch003

Chhabra, S. (2018). Framework for Enhancing Organizational Performance: Haryana Government Departments, India. In S. Chhabra (Ed.), *Handbook of Research on Civic Engagement and Social Change in Contemporary Society* (pp. 169–182). Hershey, PA: IGI Global. doi:10.4018/978-1-5225-4197-4.ch010

Criado, J. I., & Ruvalcaba-Gomez, E. A. (2018). Perceptions of City Managers About Open Government Policies: Concepts, Development, and Implementation in the Local Level of Government in Spain. *International Journal of Electronic Government Research*, *14*(1), 1–22. doi:10.4018/IJEGR.2018010101

da Rosa, I., & de Almeida, J. (2017). Digital Transformation in the Public Sector: Electronic Procurement in Portugal. In R. Shakya (Ed.), *Digital Governance and E-Government Principles Applied to Public Procurement* (pp. 99–125). Hershey, PA: IGI Global. doi:10.4018/978-1-5225-2203-4.ch005

Daoud, I. M., & Meddeb, M. (2019). The Role of Social Marketing in Preventing the Spread of Non-Communicable Diseases: Case of Tunisia. In R. Abassi & A. Ben Chehida Douss (Eds.), *Security Frameworks in Contemporary Electronic Government* (pp. 76–95). Hershey, PA: IGI Global. doi:10.4018/978-1-5225-5984-9.ch004

del Mar Gálvez-Rodríhuez, M., Haro-de-Rosario, A., & Caba-Pérez, M. D. (2017). A Comparative View of Citizen Engagement in Social Media of Local Governments from North American Countries. In M. Adria & Y. Mao (Eds.), *Handbook of Research on Citizen Engagement and Public Participation in the Era of New Media* (pp. 139–156). Hershey, PA: IGI Global. doi:10.4018/978-1-5225-1081-9.ch009

Dewani, N. D. (2019). Internet Service Provider Liability in Relation to P2P Sites: The Pirate Bay Case. In R. Abassi, & A. Ben Chehida Douss (Eds.), Security Frameworks in Contemporary Electronic Government (pp. 173-190). Hershey, PA: IGI Global. doi:10.4018/978-1-5225-5984-9.ch008

Dhamija, A., & Dhamija, D. (2018). Motivational Factors and Barriers for Technology-Driven Governance: An Indian Perspective. In A. Manoharan & J. McQuiston (Eds.), *Innovative Perspectives on Public Administration in the Digital Age* (pp. 194–211). Hershey, PA: IGI Global. doi:10.4018/978-1-5225-5966-5.ch010

Diirr, B., Araujo, R., & Cappelli, C. (2019). Empowering Society Participation in Public Service Processes. In A. Molnar (Ed.), *Strategic Management and Innovative Applications of E-Government* (pp. 145–175). Hershey, PA: IGI Global. doi:10.4018/978-1-5225-6204-7.ch007

Drulă, G. (2017). Citizen Engagement and News Selection for Facebook Pages. In M. Adria & Y. Mao (Eds.), *Handbook of Research on Citizen Engagement and Public Participation in the Era of New Media* (pp. 122–138). Hershey, PA: IGI Global. doi:10.4018/978-1-5225-1081-9.ch008

During, R., Pleijte, M., van Dam, R. I., & Salverda, I. E. (2017). The Dutch Participation Society Needs Open Data, but What is Meant by Open? In M. Adria & Y. Mao (Eds.), *Handbook of Research on Citizen Engagement and Public Participation in the Era of New Media* (pp. 304–322). Hershey, PA: IGI Global. doi:10.4018/978-1-5225-1081-9.ch017

Elena, S., & van Schalkwyk, F. (2017). Open Data for Open Justice in Seven Latin American Countries. In C. Jiménez-Gómez & M. Gascó-Hernández (Eds.), *Achieving Open Justice through Citizen Participation and Transparency* (pp. 210–231). Hershey, PA: IGI Global. doi:10.4018/978-1-5225-0717-8.ch011

Erdenebold, T. (2017). A Smart Government Framework for Mobile Application Services in Mongolia. In S. Zoughbi (Ed.), *Securing Government Information and Data in Developing Countries* (pp. 90–103). Hershey, PA: IGI Global. doi:10.4018/978-1-5225-1703-0.ch005

Erskine, M. A., & Pepper, W. (2019). Toward the Improvement of Emergency Response Utilizing a Multi-Tiered Systems Integration Approach: A Research Framework. In A. Molnar (Ed.), *Strategic Management and Innovative Applications of E-Government* (pp. 1–25). Hershey, PA: IGI Global. doi:10.4018/978-1-5225-6204-7.ch001

Essien, E. (2018). Strengthening Performance of Civil Society Through Dialogue and Critical Thinking in Nigeria: Its Ethical Implications. In S. Chhabra (Ed.), *Handbook of Research on Civic Engagement and Social Change in Contemporary Society* (pp. 82–102). Hershey, PA: IGI Global. doi:10.4018/978-1-5225-4197-4.ch005

Fan, Q. (2018). Developing A Model for Transforming Government in the Digital Age: Local Digital Government in Australia. *International Journal of E-Entrepreneurship and Innovation*, 8(2), 44–53. doi:10.4018/IJEEI.2018070104

Gascó-Hernández, M. (2017). Digitalizing Police Requirements: Opening up Justice through Collaborative Initiatives. In C. Jiménez-Gómez & M. Gascó-Hernández (Eds.), *Achieving Open Justice through Citizen Participation and Transparency* (pp. 157–172). Hershey, PA: IGI Global. doi:10.4018/978-1-5225-0717-8.ch008

Gechlik, M., Dai, D., & Beck, J. C. (2017). Open Judiciary in a Closed Society: A Paradox in China? In C. Jiménez-Gómez & M. Gascó-Hernández (Eds.), *Achieving Open Justice through Citizen Participation and Transparency* (pp. 56–92). Hershey, PA: IGI Global. doi:10.4018/978-1-5225-0717-8.ch004

Ghany, K. K., & Zawbaa, H. M. (2017). Hybrid Biometrics and Watermarking Authentication. In S. Zoughbi (Ed.), *Securing Government Information and Data in Developing Countries* (pp. 37–61). Hershey, PA: IGI Global. doi:10.4018/978-1-5225-1703-0.ch003

Gow, G. A. (2020). Alternative Social Media for Outreach and Engagement: Considering Technology Stewardship as a Pathway to Adoption. In M. Adria (Ed.), *Using New Media for Citizen Engagement and Participation* (pp. 160–180). Hershey, PA: IGI Global. doi:10.4018/978-1-7998-1828-1.ch009

Groshek, J. (2017). Organically Modified News Networks: Gatekeeping in Social Media Coverage of GMOs. In M. Adria & Y. Mao (Eds.), *Handbook of Research on Citizen Engagement and Public Participation in the Era of New Media* (pp. 107–121). Hershey, PA: IGI Global. doi:10.4018/978-1-5225-1081-9.ch007

Groshek, J. (2020). Exploring a Methodological Model for Social Media Gatekeeping on Contentious Topics: A Case Study of Twitter Interactions About GMOs. In M. Adria (Ed.), *Using New Media for Citizen Engagement and Participation* (pp. 97–111). Hershey, PA: IGI Global. doi:10.4018/978-1-7998-1828-1.ch006

Gyamfi, G. D., Gyan, G., Ayebea, M., Nortey, F. N., & Baidoo, P. Y. (2019). Assessing the Factors Affecting the Implementation of E-Government and Effect on Performance of DVLA. *International Journal of Electronic Government Research*, *15*(1), 47–61. doi:10.4018/IJEGR.2019010104

Haahr, L. (2019). Facebook Contradictions in Municipal Social Media Practices. In A. Molnar (Ed.), *Strategic Management and Innovative Applications of E-Government* (pp. 51–71). Hershey, PA: IGI Global. doi:10.4018/978-1-5225-6204-7.ch003

Hamdane, B., & El Fatmi, S. G. (2019). Information-Centric Networking, E-Government, and Security. In R. Abassi & A. Ben Chehida Douss (Eds.), *Security Frameworks in Contemporary Electronic Government* (pp. 51–75). Hershey, PA: IGI Global. doi:10.4018/978-1-5225-5984-9.ch003

Hansson, K., & Ekenberg, L. (2018). Embodiment and Gameplay: Situating the User in Crowdsourced Information Production. In A. Manoharan & J. McQuiston (Eds.), *Innovative Perspectives on Public Administration in the Digital Age* (pp. 239–255). Hershey, PA: IGI Global. doi:10.4018/978-1-5225-5966-5.ch013

Harper, L., & Sanchez, D. (2017). Electronic Government Procurement in Latin America and the Caribbean. In R. Shakya (Ed.), *Digital Governance and E-Government Principles Applied to Public Procurement* (pp. 203–228). Hershey, PA: IGI Global. doi:10.4018/978-1-5225-2203-4.ch009

Hasan, M. M., Anagnostopoulos, D., Kousiouris, G., Stamati, T., Loucopoulos, P., & Nikolaidou, M. (2019). An Ontology based Framework for E-Government Regulatory Requirements Compliance. *International Journal of E-Services and Mobile Applications*, *11*(2), 22–42. doi:10.4018/IJESMA.2019040102

Heaton, L., & Días da Silva, P. (2017). Citizen Engagement in Local Environmental Issues: Intersecting Modes of Communication. In M. Adria & Y. Mao (Eds.), *Handbook of Research on Citizen Engagement and Public Participation in the Era of New Media* (pp. 1–19). Hershey, PA: IGI Global. doi:10.4018/978-1-5225-1081-9.ch001

Heaton, L., & Días da Silva, P. (2020). Civic Engagement in Local Environmental Initiatives: Reaping the Benefits of a Diverse Media Landscape. In M. Adria (Ed.), *Using New Media for Citizen Engagement and Participation* (pp. 16–34). Hershey, PA: IGI Global. doi:10.4018/978-1-7998-1828-1.ch002

Hebbar, S., & Kiran, K. B. (2019). Social Media Influence and Mobile Government Adoption: A Conceptual Framework and its Validation. *International Journal of Electronic Government Research*, *15*(3), 37–58. doi:10.4018/IJEGR.2019070103

Heisler, E. (2017). Measuring Enclosures and Efficacy in Online Feminism: The Case of Rewire. In M. Adria & Y. Mao (Eds.), *Handbook of Research on Citizen Engagement and Public Participation in the Era of New Media* (pp. 389–409). Hershey, PA: IGI Global. doi:10.4018/978-1-5225-1081-9.ch021

Hu, F., & Yang, J. (2020). A Fuzzy Performance Evaluation Model for Government Websites Based on Language Property and Balanced Score Card. *International Journal of Enterprise Information Systems*, *16*(2), 148–163. doi:10.4018/IJEIS.2020040109

Hucks, D., Sturtz, T., & Tirabassi, K. (2018). Building, Shaping, and Modeling Tools for Literacy Development and Civic Engagement. In *Fostering Positive Civic Engagement Among Millennials: Emerging Research and Opportunities* (pp. 22–35). Hershey, PA: IGI Global. doi:10.4018/978-1-5225-2452-6.ch002

Hucks, D., Sturtz, T., & Tirabassi, K. (2018). The Role of Mentoring in Promoting Civic Engagement. In *Fostering Positive Civic Engagement Among Millennials: Emerging Research and Opportunities* (pp. 36–41). Hershey, PA: IGI Global. doi:10.4018/978-1-5225-2452-6.ch003

Husain, M. S., & Khan, N. (2017). Big Data on E-Government. In S. Zoughbi (Ed.), *Securing Government Information and Data in Developing Countries* (pp. 27–36). Hershey, PA: IGI Global. doi:10.4018/978-1-5225-1703-0.ch002

Husain, M. S., & Khanum, M. A. (2017). Cloud Computing in E-Governance: Indian Perspective. In S. Zoughbi (Ed.), *Securing Government Information and Data in Developing Countries* (pp. 104–114). Hershey, PA: IGI Global. doi:10.4018/978-1-5225-1703-0.ch006

Idoughi, D., & Abdelhakim, D. (2018). Developing Countries E-Government Services Evaluation Identifying and Testing Antecedents of Satisfaction Case of Algeria. *International Journal of Electronic Government Research*, *14*(1), 63–85. doi:10.4018/IJEGR.2018010104

Ihebuzor, N., & Egbunike, N. A. (2018). Fencism: An Unusual Political Alignment in Twitter Nigeria. In S. Chhabra (Ed.), *Handbook of Research on Civic Engagement and Social Change in Contemporary Society* (pp. 364–382). Hershey, PA: IGI Global. doi:10.4018/978-1-5225-4197-4.ch021

Ilhan, N., & Rahim, M. M. (2017). Benefits of E-Procurement Systems Implementation: Experience of an Australian Municipal Council. In R. Shakya (Ed.), *Digital Governance and E-Government Principles Applied to Public Procurement* (pp. 249–266). Hershey, PA: IGI Global. doi:10.4018/978-1-5225-2203-4.ch012

Iranmanesh, H., Keramati, A., & Behmanesh, I. (2019). The Effect of Service Innovation on E-government Performance: The Role of Stakeholders and Their Perceived Value of Innovation. *International Journal of Information Systems and Social Change*, *10*(1), 1–22. doi:10.4018/IJISSC.2019010101

Iskandarli, G. Y. (2020). Using Hotspot Information to Evaluate Citizen Satisfaction in E-Government: Hotspot Information. *International Journal of Public Administration in the Digital Age*, *7*(1), 47–62. doi:10.4018/IJPADA.2020010104

Ismailova, R., Muhametjanova, G., & Kurambayev, B. (2018). A Central Asian View of E-Government Services Adoption: Citizens' Trust and Intention to Use E-Services. In A. Manoharan & J. McQuiston (Eds.), *Innovative Perspectives on Public Administration in the Digital Age* (pp. 212–226). Hershey, PA: IGI Global. doi:10.4018/978-1-5225-5966-5.ch011

Iwasaki, Y. (2017). Youth Engagement in the Era of New Media. In M. Adria & Y. Mao (Eds.), *Handbook of Research on Citizen Engagement and Public Participation in the Era of New Media* (pp. 90–105). Hershey, PA: IGI Global. doi:10.4018/978-1-5225-1081-9.ch006

Iwasaki, Y. (2020). Centrality of Youth Engagement in Media Involvement. In M. Adria (Ed.), *Using New Media for Citizen Engagement and Participation* (pp. 81–96). Hershey, PA: IGI Global. doi:10.4018/978-1-7998-1828-1.ch005

Ji, Q. (2017). A Comparative View of Censored and Uncensored Political Discussion: The Case of Chinese Social Media Users. In M. Adria & Y. Mao (Eds.), *Handbook of Research on Citizen Engagement and Public Participation in the Era of New Media* (pp. 269–282). Hershey, PA: IGI Global. doi:10.4018/978-1-5225-1081-9.ch015

Jiménez-Gómez, C. E. (2017). Open Judiciary Worldwide: Best Practices and Lessons Learnt. In C. Jiménez-Gómez & M. Gascó-Hernández (Eds.), *Achieving Open Justice through Citizen Participation and Transparency* (pp. 1–15). Hershey, PA: IGI Global. doi:10.4018/978-1-5225-0717-8.ch001

Karamagioli, E., Staiou, E. R., & Gouscos, D. (2018). Assessing the Social Media Presence and Activity of Major Greek Cities During 2014-2017: Towards Local Government 2.0? In S. Chhabra (Ed.), *Handbook of Research on Civic Engagement and Social Change in Contemporary Society* (pp. 272–293). Hershey, PA: IGI Global. doi:10.4018/978-1-5225-4197-4.ch016

Kasemsap, K. (2017). Mastering Electronic Procurement, Green Public Procurement, and Public Procurement for Innovation. In R. Shakya (Ed.), *Digital Governance and E-Government Principles Applied to Public Procurement* (pp. 29–55). Hershey, PA: IGI Global. doi:10.4018/978-1-5225-2203-4.ch002

Kassen, M. (2017). E-Government Politics as a Networking Phenomenon: Applying a Multidimensional Approach. *International Journal of Electronic Government Research*, *13*(2), 18–46. doi:10.4018/IJEGR.2017040102

Kaya, T., Sağsan, M., Yıldız, M., Medeni, T., & Medeni, T. (2020). Citizen Attitudes Towards E-Government Services: Comparison of Northern and Southern Nicosia Municipalities. *International Journal of Public Administration in the Digital Age*, *7*(1), 17–32. doi:10.4018/IJPADA.2020010102

Khan, M. I., Foley, S. N., & O'Sullivan, B. (2019). DBMS Log Analytics for Detecting Insider Threats in Contemporary Organizations. In R. Abassi & A. Ben Chehida Douss (Eds.), *Security Frameworks in Contemporary Electronic Government* (pp. 207–234). Hershey, PA: IGI Global. doi:10.4018/978-1-5225-5984-9.ch010

Kilhoffer, Z. (2020). Platform Work and Participation: Disentangling the Rhetoric. In M. Adria (Ed.), *Using New Media for Citizen Engagement and Participation* (pp. 1–15). Hershey, PA: IGI Global. doi:10.4018/978-1-7998-1828-1.ch001

Kita, Y. (2017). An Analysis of a Lay Adjudication System and Open Judiciary: The New Japanese Lay Adjudication System. In C. Jiménez-Gómez & M. Gascó-Hernández (Eds.), *Achieving Open Justice through Citizen Participation and Transparency* (pp. 93–109). Hershey, PA: IGI Global. doi:10.4018/978-1-5225-0717-8.ch005

Lacharite, J. R. (2017). Digital Media, Civic Literacy, and Civic Engagement: The "Promise and Peril" of Internet Politics in Canada. In R. Luppicini & R. Baarda (Eds.), *Digital Media Integration for Participatory Democracy* (pp. 44–65). Hershey, PA: IGI Global. doi:10.4018/978-1-5225-2463-2.ch003

Lampoltshammer, T. J., Guadamuz, A., Wass, C., & Heistracher, T. (2017). Openlaws. eu: Open Justice in Europe through Open Access to Legal Information. In C. Jiménez-Gómez & M. Gascó-Hernández (Eds.), *Achieving Open Justice through Citizen Participation and Transparency* (pp. 173–190). Hershey, PA: IGI Global. doi:10.4018/978-1-5225-0717-8.ch009

Leocadia, D. R. (2018). The Use of Social Media by Local Governments: Benefits, Challenges, and Recent Experiences. In S. Chhabra (Ed.), *Handbook of Research on Civic Engagement and Social Change in Contemporary Society* (pp. 294–310). Hershey, PA: IGI Global. doi:10.4018/978-1-5225-4197-4.ch017

Liao, Y. (2018). How to Foster Citizen Reblogging of a Government Microblog: Evidence From Local Government Publicity Microblogs in China. *International Journal of Public Administration in the Digital Age*, 5(3), 1–15. doi:10.4018/IJPADA.2018070101

Lourenço, R. P., Fernando, P., & Gomes, C. (2017). From eJustice to Open Judiciary: An Analysis of the Portuguese Experience. In C. Jiménez-Gómez & M. Gascó-Hernández (Eds.), *Achieving Open Justice through Citizen Participation and Transparency* (pp. 111–136). Hershey, PA: IGI Global. doi:10.4018/978-1-5225-0717-8.ch006

Luk, C. Y. (2018). The Impact of Digital Health on Traditional Healthcare Systems and Doctor-Patient Relationships: The Case Study of Singapore. In A. Manoharan & J. McQuiston (Eds.), *Innovative Perspectives on Public Administration in the Digital Age* (pp. 143–167). Hershey, PA: IGI Global. doi:10.4018/978-1-5225-5966-5.ch008

Luk, C. Y. (2019). Strengthening Cybersecurity in Singapore: Challenges, Responses, and the Way Forward. In R. Abassi & A. Ben Chehida Douss (Eds.), *Security Frameworks in Contemporary Electronic Government* (pp. 96–128). Hershey, PA: IGI Global. doi:10.4018/978-1-5225-5984-9.ch005

Luppicini, R. (2017). Technoethics and Digital Democracy for Future Citizens. In R. Luppicini & R. Baarda (Eds.), *Digital Media Integration for Participatory Democracy* (pp. 1–21). Hershey, PA: IGI Global. doi:10.4018/978-1-5225-2463-2.ch001

Ma, L. (2018). Digital Divide and Citizen Use of E-Government in China's Municipalities. *International Journal of Public Administration in the Digital Age*, 5(3), 16–31. doi:10.4018/IJPADA.2018070102

Madsen, C. Ø., & Kræmmergaard, P. (2018). How to Migrate Citizens Online and Reduce Traffic on Traditional Channels Through Multichannel Management: A Case Study of Cross-Organizational Collaboration Surrounding a Mandatory Self-Service Application. In A. Manoharan & J. McQuiston (Eds.), *Innovative Perspectives on Public Administration in the Digital Age* (pp. 121–142). Hershey, PA: IGI Global. doi:10.4018/978-1-5225-5966-5.ch007

Mahmood, M. (2019). Transformation of Government and Citizen Trust in Government: A Conceptual Model. In A. Molnar (Ed.), *Strategic Management and Innovative Applications of E-Government* (pp. 107–122). Hershey, PA: IGI Global. doi:10.4018/978-1-5225-6204-7.ch005

Marcovecchio, I., Thinyane, M., Estevez, E., & Janowski, T. (2019). Digital Government as Implementation Means for Sustainable Development Goals. *International Journal of Public Administration in the Digital Age*, 6(3), 1–22. doi:10.4018/IJPADA.2019070101

Marques, B., McIntosh, J., & Campays, P. (2018). Participatory Design for Under-Represented Communities: A Collaborative Design-Led Research Approach for Place-Making. In S. Chhabra (Ed.), *Handbook of Research on Civic Engagement and Social Change in Contemporary Society* (pp. 1–15). Hershey, PA: IGI Global. doi:10.4018/978-1-5225-4197-4.ch001

Matiatou, M. (2018). NGOs, Civil Society, and Global Governance in the Era of Sustainability and Consolidation: A Taxonomy of Value. In S. Chhabra (Ed.), *Handbook of Research on Civic Engagement and Social Change in Contemporary Society* (pp. 183–206). Hershey, PA: IGI Global. doi:10.4018/978-1-5225-4197-4.ch011

Maulana, I. (2020). Social Media as Public Political Instrument. In M. Adria (Ed.), *Using New Media for Citizen Engagement and Participation* (pp. 181–197). Hershey, PA: IGI Global. doi:10.4018/978-1-7998-1828-1.ch010

Maxwell, S., & Carboni, J. (2018). Foundations' Civic Engagement With Stakeholders: Leveraging Social Media Strategies. In A. Manoharan & J. McQuiston (Eds.), *Innovative Perspectives on Public Administration in the Digital Age* (pp. 25–41). Hershey, PA: IGI Global. doi:10.4018/978-1-5225-5966-5.ch002

McCarthy-Latimer, C. (2018). Higher Education Pedagogy Revisited: Impacting Political Science College Students' Active Learning, Opinion Development, and Participation. In S. Chhabra (Ed.), *Handbook of Research on Civic Engagement and Social Change in Contemporary Society* (pp. 63–81). Hershey, PA: IGI Global. doi:10.4018/978-1-5225-4197-4.ch004

McNeal, R. S., Schmeida, M., & Bryan, L. D. (2017). Smartphones and their Increased Importance in U.S. Presidential Elections. In M. Adria & Y. Mao (Eds.), *Handbook of Research on Citizen Engagement and Public Participation in the Era of New Media* (pp. 283–303). Hershey, PA: IGI Global. doi:10.4018/978-1-5225-1081-9.ch016

Mengesha, N., Ayanso, A., & Demissie, D. (2020). Profiles and Evolution of E-Government Readiness in Africa: A Segmentation Analysis. *International Journal of Information Systems and Social Change*, *11*(1), 43–65. doi:10.4018/IJISSC.2020010104

Mensah, I. K. (2019). Exploring the Moderating Effect of Perceived Usefulness on the Adoption of E-Government Services. *International Journal of Electronic Government Research*, *15*(1), 17–35. doi:10.4018/IJEGR.2019010102

Mensah, I. K., & Mi, J. (2018). An Empirical Investigation of the Impact of Demographic Factors on E-Government Services Adoption. *International Journal of E-Services and Mobile Applications*, *10*(2), 17–35. doi:10.4018/IJESMA.2018040102

Mensah, I. K., & Mi, J. (2018). Determinants of Intention to Use Local E-Government Services in Ghana: The Perspective of Local Government Workers. *International Journal of Technology Diffusion*, *9*(2), 41–60. doi:10.4018/IJTD.2018040103

Mensah, I. K., & Mi, J. (2019). Predictors of the Readiness to Use E-Government Services From Citizens' Perspective. *International Journal of Technology Diffusion*, *10*(1), 39–59. doi:10.4018/IJTD.2019010103

Mensah, I. K., Vera, P., & Mi, J. (2018). Factors Determining the Use of E-Government Services: An Empirical Study on Russian Students in China. *International Journal of E-Adoption*, *10*(2), 1–19. doi:10.4018/IJEA.2018070101

Merhi, M. I. (2018). Does National Culture Have Any Impact on E-Government Usage? *International Journal of Technology Diffusion*, *9*(3), 29–45. doi:10.4018/IJTD.2018070103

Metelmann, C., & Metelmann, B. (2019). Live Video Communication in Prehospital Emergency Medicine. In A. Molnar (Ed.), *Strategic Management and Innovative Applications of E-Government* (pp. 26–50). Hershey, PA: IGI Global. doi:10.4018/978-1-5225-6204-7.ch002

Mohammed-Spigner, D., Porter, B. E., & Warner, L. M. (2018). States' Budget Investments in Technology and Improving Criminal Justice Outcomes. In A. Manoharan & J. McQuiston (Eds.), *Innovative Perspectives on Public Administration in the Digital Age* (pp. 90–105). Hershey, PA: IGI Global. doi:10.4018/978-1-5225-5966-5.ch005

Moreno, L., & Martínez, P. (2019). Accessibility Compliance for E-Government Websites: Laws, Standards, and Evaluation Technology. *International Journal of Electronic Government Research*, *15*(2), 1–18. doi:10.4018/IJEGR.2019040101

Mwirigi, G. B., Zo, H., Rho, J. J., & Park, M. J. (2017). An Empirical Investigation of M-Government Acceptance in Developing Countries: A Case of Kenya. In S. Zoughbi (Ed.), *Securing Government Information and Data in Developing Countries* (pp. 62–89). Hershey, PA: IGI Global. doi:10.4018/978-1-5225-1703-0.ch004

Naeem, M. (2019). Uncovering the Enablers, Benefits, Opportunities and Risks for Digital Open Government (DOG): Enablers, Benefits, Opportunities and Risks for DOG. *International Journal of Public Administration in the Digital Age*, *6*(3), 41–58. doi:10.4018/IJPADA.2019070103

Naidoo, V., & Nzimakwe, T. I. (2019). M-Government and Its Application on Public Service Delivery. In R. Abassi & A. Ben Chehida Douss (Eds.), *Security Frameworks in Contemporary Electronic Government* (pp. 1–14). Hershey, PA: IGI Global. doi:10.4018/978-1-5225-5984-9.ch001

Nam, T. (2017). Achievable or Ambitious?: A Comparative and Critical View of Government 3.0 in Korea. *International Journal of Electronic Government Research*, *13*(1), 1–13. doi:10.4018/IJEGR.2017010101

Nam, T. (2019). Does E-Government Raise Effectiveness and Efficiency?: Examining the Cross-National Effect. *Journal of Global Information Management, 27*(3), 120–138. doi:10.4018/JGIM.2019070107

Nasri, W. (2019). E-Government Adoption in Tunisia Extending Technology Acceptance Model. *International Journal of Public Administration in the Digital Age, 6*(4), 30–42. doi:10.4018/IJPADA.2019100103

Nathai-Balkissoon, M., & Pun, K. F. (2018). E-Government of Occupational Safety and Health: Improvement Prospects for a Developing Nation. In A. Manoharan & J. McQuiston (Eds.), *Innovative Perspectives on Public Administration in the Digital Age* (pp. 65–89). Hershey, PA: IGI Global. doi:10.4018/978-1-5225-5966-5.ch004

Nayak, S. (2018). Agriculture Livelihood Security: Industry CSR Initiative. In S. Chhabra (Ed.), *Handbook of Research on Civic Engagement and Social Change in Contemporary Society* (pp. 103–118). Hershey, PA: IGI Global. doi:10.4018/978-1-5225-4197-4.ch006

Neupane, A., Soar, J., Vaidya, K., & Aryal, S. (2017). Application of E-Government Principles in Anti-Corruption Framework. In R. Shakya (Ed.), *Digital Governance and E-Government Principles Applied to Public Procurement* (pp. 56–74). Hershey, PA: IGI Global. doi:10.4018/978-1-5225-2203-4.ch003

Niewiadomski, R., & Anderson, D. (2018). Saving Democracy: Populism, Deception, and E-Participation. In L. Alcaide-Muñoz & F. Alcaraz-Quiles (Eds.), *Optimizing E-Participation Initiatives Through Social Media* (pp. 54–75). Hershey, PA: IGI Global. doi:10.4018/978-1-5225-5326-7.ch003

Noor, R. (2020). Citizen Journalism: New-Age Newsgathering. In M. Adria (Ed.), *Using New Media for Citizen Engagement and Participation* (pp. 135–159). Hershey, PA: IGI Global. doi:10.4018/978-1-7998-1828-1.ch008

Nurdin, N. (2018). Institutional Arrangements in E-Government Implementation and Use: A Case Study From Indonesian Local Government. *International Journal of Electronic Government Research, 14*(2), 44–63. doi:10.4018/IJEGR.2018040104

Okada, A., Ishida, Y., & Yamauchi, N. (2018). In Prosperity Prepare for Adversity: Use of Social Media for Nonprofit Fundraising in Times of Disaster. In A. Manoharan, & J. McQuiston (Eds.), Innovative Perspectives on Public Administration in the Digital Age (pp. 42-64). Hershey, PA: IGI Global. doi:10.4018/978-1-5225-5966-5.ch003

Omar, A., Johnson, C., & Weerakkody, V. (2019). Debating Digitally-Enabled Service Transformation in Public Sector: Keeping the Research Talking. In A. Molnar (Ed.), *Strategic Management and Innovative Applications of E-Government* (pp. 123–144). Hershey, PA: IGI Global. doi:10.4018/978-1-5225-6204-7.ch006

Omotosho, B. J. (2018). Reconnoitering the Fight Against Political Corruption and the Way Forward in Nigeria. In S. Chhabra (Ed.), *Handbook of Research on Civic Engagement and Social Change in Contemporary Society* (pp. 207–219). Hershey, PA: IGI Global. doi:10.4018/978-1-5225-4197-4.ch012

Ona, S. E., & Concepcion, M. B. (2018). Building Performance Competencies in Open Government: Perspectives From the Philippines. In L. Alcaide-Muñoz & F. Alcaraz-Quiles (Eds.), *Optimizing E-Participation Initiatives Through Social Media* (pp. 176–207). Hershey, PA: IGI Global. doi:10.4018/978-1-5225-5326-7.ch008

Oravec, J. A. (2018). "Don't Be Evil" and Beyond for High Tech Organizations: Ethical Statements and Mottos (and Responsibility). In S. Chhabra (Ed.), *Handbook of Research on Civic Engagement and Social Change in Contemporary Society* (pp. 220–237). Hershey, PA: IGI Global. doi:10.4018/978-1-5225-4197-4.ch013

Ortuzar, G. B., Sevillano, E. M., Castro, C. L., & Uribe, C. (2017). Challenges in Chilean E-Procurement System: A Critical Review. In R. Shakya (Ed.), *Digital Governance and E-Government Principles Applied to Public Procurement* (pp. 170–202). Hershey, PA: IGI Global. doi:10.4018/978-1-5225-2203-4.ch008

Panda, P., & Sahu, G. P. (2017). Public Procurement Framework in India: An Overview. In R. Shakya (Ed.), *Digital Governance and E-Government Principles Applied to Public Procurement* (pp. 229–248). Hershey, PA: IGI Global. doi:10.4018/978-1-5225-2203-4.ch010

Perelló-Sobrepere, M. (2017). Building a New State from Outrage: The Case of Catalonia. In M. Adria & Y. Mao (Eds.), *Handbook of Research on Citizen Engagement and Public Participation in the Era of New Media* (pp. 344–359). Hershey, PA: IGI Global. doi:10.4018/978-1-5225-1081-9.ch019

Quental, C., & Gouveia, L. B. (2018). Participation Sphere: A Model and a Framework for Fostering Participation in Organizations. In S. Chhabra (Ed.), *Handbook of Research on Civic Engagement and Social Change in Contemporary Society* (pp. 16–39). Hershey, PA: IGI Global. doi:10.4018/978-1-5225-4197-4.ch002

Raaen, N. (2017). Open and Transparent Judicial Records in the Digital Age: Applying Principles and Performance Measures. In C. Jiménez-Gómez & M. Gascó-Hernández (Eds.), *Achieving Open Justice through Citizen Participation and Transparency* (pp. 16–35). Hershey, PA: IGI Global. doi:10.4018/978-1-5225-0717-8.ch002

Raina, R. L., Alam, I., & Siddiqui, F. (2018). Facebook Losing to WhatsApp: The Changing Social Networking Patterns in India. In S. Chhabra (Ed.), *Handbook of Research on Civic Engagement and Social Change in Contemporary Society* (pp. 328–346). Hershey, PA: IGI Global. doi:10.4018/978-1-5225-4197-4.ch019

Randa, I. O. (2018). Leveraging Hybrid Value Chain for Affordable Housing Delivery in the City of Windhoek. In S. Chhabra (Ed.), *Handbook of Research on Civic Engagement and Social Change in Contemporary Society* (pp. 119–141). Hershey, PA: IGI Global. doi:10.4018/978-1-5225-4197-4.ch007

Roy, J. P. (2019). Service, Openness and Engagement as Digitally-Based Enablers of Public Value?: A Critical Examination of Digital Government in Canada. *International Journal of Public Administration in the Digital Age, 6*(3), 23–40. doi:10.4018/IJPADA.2019070102

Saidane, A., & Al-Sharieh, S. (2019). A Compliance-Driven Framework for Privacy and Security in Highly Regulated Socio-Technical Environments: An E-Government Case Study. In R. Abassi & A. Ben Chehida Douss (Eds.), *Security Frameworks in Contemporary Electronic Government* (pp. 15–50). Hershey, PA: IGI Global. doi:10.4018/978-1-5225-5984-9.ch002

Salvati, E. (2017). E-Government and E-Democracy in the Supranational Arena: The Enforcing of Transparency and Democratic Legitimacy in the European Union. In R. Luppicini & R. Baarda (Eds.), *Digital Media Integration for Participatory Democracy* (pp. 101–129). Hershey, PA: IGI Global. doi:10.4018/978-1-5225-2463-2.ch006

Sandoval-Almazán, R. (2017). Open Justice in Latin America?: An Assessment Framework for Judiciary Portals in 2015. In C. Jiménez-Gómez & M. Gascó-Hernández (Eds.), *Achieving Open Justice through Citizen Participation and Transparency* (pp. 232–252). Hershey, PA: IGI Global. doi:10.4018/978-1-5225-0717-8.ch012

Santos, H. R., & Tonelli, D. F. (2019). Smart Government and the Maturity Levels of Sociopolitical Digital Interactions: Analysing Temporal Changes in Brazilian E-Government Portals. In A. Molnar (Ed.), *Strategic Management and Innovative Applications of E-Government* (pp. 176–199). Hershey, PA: IGI Global. doi:10.4018/978-1-5225-6204-7.ch008

Sayogo, D. S., & Yuli, S. B. (2018). Critical Success Factors of Open Government and Open Data at Local Government Level in Indonesia. *International Journal of Electronic Government Research, 14*(2), 28–43. doi:10.4018/IJEGR.2018040103

Scholl, H. J. (2019). Strategic Overhaul of Government Operations: Situated Action Analysis of Socio-Technical Innovation in the Public Sector. In A. Molnar (Ed.), *Strategic Management and Innovative Applications of E-Government* (pp. 72–106). Hershey, PA: IGI Global. doi:10.4018/978-1-5225-6204-7.ch004

Semali, L. M. (2018). Indigenous Knowledge as Resource to Sustain Self-Employment in Rural Development. In S. Chhabra (Ed.), *Handbook of Research on Civic Engagement and Social Change in Contemporary Society* (pp. 142–158). Hershey, PA: IGI Global. doi:10.4018/978-1-5225-4197-4.ch008

Shakya, R. K., & Schapper, P. R. (2017). Digital Governance and E-Government Principles: E-Procurement as Transformative. In R. Shakya (Ed.), *Digital Governance and E-Government Principles Applied to Public Procurement* (pp. 1–28). Hershey, PA: IGI Global. doi:10.4018/978-1-5225-2203-4.ch001

Shan, Z., & Tang, L. (2020). Social Media and Public Sphere in China: A Case Study of Political Discussion on Weibo After the Wenzhou High- Speed Rail Derailment Accident. In M. Adria (Ed.), *Using New Media for Citizen Engagement and Participation* (pp. 280–295). Hershey, PA: IGI Global. doi:10.4018/978-1-7998-1828-1.ch015

Sharma, V., & Sengar, A. (2018). Mobile Governance for Affordability and Profitability/Feasibility. In A. Manoharan & J. McQuiston (Eds.), *Innovative Perspectives on Public Administration in the Digital Age* (pp. 227–238). Hershey, PA: IGI Global. doi:10.4018/978-1-5225-5966-5.ch012

Sihi, D. (2018). Friending and Funding Through Facebook: Social Media Use of Regional Nonprofit Organizations. In A. Manoharan & J. McQuiston (Eds.), *Innovative Perspectives on Public Administration in the Digital Age* (pp. 256–283). Hershey, PA: IGI Global. doi:10.4018/978-1-5225-5966-5.ch014

Soares de Freitas, C., & Ewerton, I. N. (2018). Networks for Cyberactivism and Their Implications for Policymaking in Brazil. In L. Alcaide-Muñoz & F. Alcaraz-Quiles (Eds.), *Optimizing E-Participation Initiatives Through Social Media* (pp. 155–175). Hershey, PA: IGI Global. doi:10.4018/978-1-5225-5326-7.ch007

Song, M. Y. (2020). Public Engagement and Policy Entrepreneurship on Social Media in the Time of Anti-Vaccination Movements. In M. Adria (Ed.), *Using New Media for Citizen Engagement and Participation* (pp. 60–80). Hershey, PA: IGI Global. doi:10.4018/978-1-7998-1828-1.ch004

Soverchia, M., & Fradeani, A. (2018). The eXtensible Business Reporting Language: A Digital Tool to Enhance Public Sector Accountability. In A. Manoharan & J. McQuiston (Eds.), *Innovative Perspectives on Public Administration in the Digital Age* (pp. 106–120). Hershey, PA: IGI Global. doi:10.4018/978-1-5225-5966-5.ch006

Spada, P., & Allegretti, G. (2020). When Democratic Innovations Integrate Multiple and Diverse Channels of Social Dialogue: Opportunities and Challenges. In M. Adria (Ed.), *Using New Media for Citizen Engagement and Participation* (pp. 35–59). Hershey, PA: IGI Global. doi:10.4018/978-1-7998-1828-1.ch003

Stacey, E. (2018). Networked Protests: A Review of Social Movement Literature and the Hong Kong Umbrella Movement (2017). In S. Chhabra (Ed.), *Handbook of Research on Civic Engagement and Social Change in Contemporary Society* (pp. 347–363). Hershey, PA: IGI Global. doi:10.4018/978-1-5225-4197-4.ch020

Staiou, E., & Gouscos, D. (2018). The Evolution of Self-Organized Social Solidarity (SoSS) Initiatives in Greece and Their Relationship to Online Media: A Longitudinal Research. In A. Manoharan & J. McQuiston (Eds.), *Innovative Perspectives on Public Administration in the Digital Age* (pp. 168–193). Hershey, PA: IGI Global. doi:10.4018/978-1-5225-5966-5.ch009

Strekalova, Y. A., Krieger, J. L., Damiani, R. E., Kalyanaraman, S., & Wang, D. Z. (2017). Old Media, New Media, and Public Engagement with Science and Technology. In M. Adria & Y. Mao (Eds.), *Handbook of Research on Citizen Engagement and Public Participation in the Era of New Media* (pp. 57–72). Hershey, PA: IGI Global. doi:10.4018/978-1-5225-1081-9.ch004

Strongman, L. (2018). Understanding Private and Public Partnerships. In S. Chhabra (Ed.), *Handbook of Research on Civic Engagement and Social Change in Contemporary Society* (pp. 159–168). Hershey, PA: IGI Global. doi:10.4018/978-1-5225-4197-4.ch009

Thomas, M. A., & Elnagar, S. (2018). A Semantic Approach to Evaluate Web Content of Government Websites. In A. Manoharan & J. McQuiston (Eds.), *Innovative Perspectives on Public Administration in the Digital Age* (pp. 1–24). Hershey, PA: IGI Global. doi:10.4018/978-1-5225-5966-5.ch001

Toscano, J. P. (2017). Social Media and Public Participation: Opportunities, Barriers, and a New Framework. In M. Adria & Y. Mao (Eds.), *Handbook of Research on Citizen Engagement and Public Participation in the Era of New Media* (pp. 73–89). Hershey, PA: IGI Global. doi:10.4018/978-1-5225-1081-9.ch005

Towner, T. L. (2020). Information Hubs or Drains?: The Role of Online Sources in Campaign Learning. In M. Adria (Ed.), *Using New Media for Citizen Engagement and Participation* (pp. 112–134). Hershey, PA: IGI Global. doi:10.4018/978-1-7998-1828-1.ch007

Tsygankov, S., & Gasanova, E. (2017). Electronification of the Public Procurement System: A Comparative Analysis of the Experience of the Russian Federation and Ukraine. In R. Shakya (Ed.), *Digital Governance and E-Government Principles Applied to Public Procurement* (pp. 267–277). Hershey, PA: IGI Global. doi:10.4018/978-1-5225-2203-4.ch013

Vakeel, K. A., & Panigrahi, P. K. (2018). Social Media Usage in E-Government: Mediating Role of Government Participation. *Journal of Global Information Management*, *26*(1), 1–19. doi:10.4018/JGIM.2018010101

Valaei, N., Nikhashemi, S. R., Jin, H. H., & Dent, M. M. (2018). Task Technology Fit in Online Transaction Through Apps. In L. Alcaide-Muñoz, & F. Alcaraz-Quiles (Eds.), Optimizing E-Participation Initiatives Through Social Media (pp. 236-251). Hershey, PA: IGI Global. doi:10.4018/978-1-5225-5326-7.ch010

Valle-Cruz, D., & Sandoval-Almazan, R. (2018). Boosting E-Participation: The Use of Social Media in Municipalities in the State of Mexico. In L. Alcaide-Muñoz & F. Alcaraz-Quiles (Eds.), *Optimizing E-Participation Initiatives Through Social Media* (pp. 103–125). Hershey, PA: IGI Global. doi:10.4018/978-1-5225-5326-7.ch005

Vie, S., Carter, D., & Meyr, J. (2017). Occupy Rhetoric: Responding to Charges of "Slacktivism" with Digital Activism Successes. In M. Adria & Y. Mao (Eds.), *Handbook of Research on Citizen Engagement and Public Participation in the Era of New Media* (pp. 179–193). Hershey, PA: IGI Global. doi:10.4018/978-1-5225-1081-9.ch011

Vij, R. (2018). Do Women Perceive Organizational Culture Differently From Men?: A Case Study of State Bank of India. In S. Chhabra (Ed.), *Handbook of Research on Civic Engagement and Social Change in Contemporary Society* (pp. 253–271). Hershey, PA: IGI Global. doi:10.4018/978-1-5225-4197-4.ch015

Villanueva-Mansilla, E. (2020). Salience, Self-Salience, and Discursive Opportunities: An Effective Media Presence Construction Through Social Media in the Peruvian Presidential Election. In M. Adria (Ed.), *Using New Media for Citizen Engagement and Participation* (pp. 240–255). Hershey, PA: IGI Global. doi:10.4018/978-1-7998-1828-1.ch013

Warf, B. (2018). Geographies of e-Government in Europe: European e-Government. *International Journal of E-Planning Research, 7*(4), 61–77. doi:10.4018/IJEPR.2018100104

Wright, R. J., & Shiner, J. M. (2017). Managing Collaboration in E-Procurement. In R. Shakya (Ed.), *Digital Governance and E-Government Principles Applied to Public Procurement* (pp. 75–98). Hershey, PA: IGI Global. doi:10.4018/978-1-5225-2203-4.ch004

Yang, K. C., & Kang, Y. (2020). Political Mobilization Strategies in Taiwan's Sunflower Student Movement on March 18, 2014: A Text-Mining Analysis of Cross-National Media Corpus. In M. Adria (Ed.), *Using New Media for Citizen Engagement and Participation* (pp. 256–279). Hershey, PA: IGI Global. doi:10.4018/978-1-7998-1828-1.ch014

Yap, C. S., Ahmad, R., Newaz, F. T., & Mason, C. (2019). Continuous Use Intention of E-Government Portals the Perspective of Older Citizens. *International Journal of Electronic Government Research, 15*(1), 1–16. doi:10.4018/IJEGR.2019010101

Yin, Y., & Fung, A. (2017). Youth Online Cultural Participation and Bilibili: An Alternative Form of Democracy in China? In R. Luppicini & R. Baarda (Eds.), *Digital Media Integration for Participatory Democracy* (pp. 130–154). Hershey, PA: IGI Global. doi:10.4018/978-1-5225-2463-2.ch007

Yuan, Y. (2020). A Gradual Political Change?: The Agenda-Setting Effect of Online Activism in China 1994-2011. In M. Adria (Ed.), *Using New Media for Citizen Engagement and Participation* (pp. 198–218). Hershey, PA: IGI Global. doi:10.4018/978-1-7998-1828-1.ch011

Zoughbi, S. (2017). Major Issues Affecting Government Data and Information in Developing Countries. In S. Zoughbi (Ed.), *Securing Government Information and Data in Developing Countries* (pp. 115–126). Hershey, PA: IGI Global. doi:10.4018/978-1-5225-1703-0.ch007

Zoughbi, S. (2017). Major Technology Trends Affecting Government Data in Developing Countries. In S. Zoughbi (Ed.), *Securing Government Information and Data in Developing Countries* (pp. 127–135). Hershey, PA: IGI Global. doi:10.4018/978-1-5225-1703-0.ch008

About the Contributors

Manuel Pedro Rodríguez Bolívar is Full Professor of Accounting at the University of Granada. He has been a visiting professor at the Utrecht Universiteit and is currently a professor of more than 4 official master's degrees, as well as a visiting lecturer in master's degrees in other universities such as the Autonomous University of Madrid. His research interests are mainly related to the implementation of new technologies in Public Administration (e-government), open government, governance in smart cities and public sector financial sustainability. He is also the author of more than 55 papers published in 29 different international journals indexed in the Journal Citation Report and has authored more than 35 book chapters published in Kluwer Academic Publishers, Springer, Routledge, Palgrave, Taylor and Francis and IGI Global. He finally holds the position of Editor in Chief of the International Journal of Public Administration in the Digital Age (IJPADA) (Emerging SSCI) and PAIT book Series in Springer, and he is a member of the Editorial Board of Government Information Quarterly (JCR) and Informatics (Emerging SSCI).

María Elicia Cortés Cediel is an Associate Lecturer at the Faculty of Political Science and Sociology of Universidad Complutense de Madrid, where she earned a MSc in Government and Public Administration (best dissertation mention in the faculty) and is pursuing a PhD on Political Sciences, Public Administration and International Relations. She earned a BSc on Law and a BSc on Political Sciences and Public Administration at Universidad Autónoma de Madrid. Interested in new forms of citizen participation supported by electronic tools, she has co-authored research papers that analyze e-participation from human, social and technological perspectives, focusing on smart governance and e-governance in smart cities.

* * *

Debora Albu holds an MSc in Gender, Development and Globalization from the London School of Economics and Political Science, in the UK, being awarded with a Chevening scholarship. She holds a B.A. in International Relations from the

Pontifical Catholic University Rio de Janeiro (PUC-Rio). In 2015, she completed the course "Gender Equality" at the University of Oslo, in Norway. She has worked in the areas of activism, youth and public policy at Amnesty International and at University of Youth. Debora was project and communications manager at the Center for International Relations at Getulio Vargas Foundation (FGV). She writes for independent online channels and develops research in the democracy and technology, digital activism and gender studies. She is program coordinator at the Democracy and Technology area at the Institute for Technology and Society.

Cristina Alcaide-Muñoz holds a PhD from the Public University of Navarre. She held a research grant in the Department of Business Administration (Public University of Navarre) to develop the line of research based on operations management and strategic management, focusing on high-performance manufacturing organizations. Her research interests include strategic planning and the use of ICTs on private and public organizations. Moreover, she teaches financial accounting, consolidated financial statements and financial statement anaylsis at International University of La Rioja (UNIR).

Laura Alcaide-Muñoz is Associate Professor in Accounting and Finance in the Department of Financial Economy and Accounting at the University of Granada. She is interested in E-government, E-Participation and Smart Cities. She has been author of numerous articles in leading SSCI journals (Business, Economics, Information Science and Public Administration) and has written book chapters in prestigious editorials like IGI Global, Springer and Routledge-Taylor & Francis. Also, she has edited books in editorial like Springer and Palgrave McMillan.

Angel Bartolomé Muñoz de Luna joined CEU University in September 2007, as associate professor of communications. He combines intensive academic research and teaching with intense political and marketing activity. He had previously launched a Startup, chosen by Entrepreneur magazine as one of the top 50 companies of the year 2011. Bartolomé Ph.D. in Advertising Cum laude (San Pablo CEU, 2010), holds a Bachelor's Degree in Advertising and Public Relations (CEU University, 2005) a Bachelor's Degree in Philosophy (San Dámaso, 2003), Master of Design Management (EOI, 2006) and certificate CCL in leadership (2016) 5 Top Financial Times Executive Education Ranking. In the year 2011 Ángel Bartolomé was appointed Director of Marketing at Marco Aldany (400 locations in Europe and America, with revenues above € 120 million, more than 6.5 million visitors per year and 4,000 employees) being responsable for the brand in Spain and Latinoamérica. In two years he headed the change the change of prices (because the increase of taxes from 7 to 21%), he modernized the country's brand image and promoted a strategy to take the company into on line markets (case study in Facebook Awards TNS, 2014) Since March 2013, he managed the Responsible for institutional communication and policy in Las Rozas Local Council. Las Rozas is one of the larger townships and municipalities in the autonomous community of Madrid with a great pull of enterprises: HP, Kyocera, Foster Wheeler, Bankia, LG, Alpahabet.

Diego Cerqueira, Software Engineer and Researcher, holds a degree in Information Systems, currently pursuing a Master's degree in Systems Engineering and Computers Science (PESC) at the Federal University of Rio de Janeiro - UFRJ. Currently Tech Leader at Mudamos, a civic participation framework winner of the Google Impact Challenge (2016). He believes in good use of technology, have been working with Civic Tech and GovTech for the past three years at the Institute for Technology and Society of Rio (ITS Rio) at the Democracy and Technology area.

Lauri Goldkind has a longstanding interest and practice background in nonprofit leadership, capacity building, and organizational development. At Fordham she teaches across the foundation and advanced years. Her practice experience has been centered in the youth development, education, and juvenile justice realms. Prior to joining the faculty, she served as the Director of New School Development and the Director of Evaluation at The Urban Assembly (UA), a network of new specialized public schools located in the Brooklyn, the Bronx and Manhattan. At UA she supported principals through the new school process, helping them earn start-up grants valued at over $500,000 per school; additionally, she provided technical assistance to principals and school-based staff on data-driven decision making, development and maintenance of data management structures and the effective use of data to improve student achievement. She has had the privilege of working with youth in NYC at organizations such as CASES, the Posse Foundation and the DOME Project. Dr. Goldkind holds an M.S.W. from SUNY Stony Brook with a concentration in planning, administration, and research and a PhD from the Wurzweiler School of Social Work at Yeshiva University.

Thayane Guimarães holds a degree in Journalism from the Fluminense Federal University and she was awarded for the special report "Underreporting of Homophobic Crimes: The Invisible Face of Violence". She has already worked on Cultural Group AfroReggae and she has integrated the Communications and Campaigns team at Amnesty International Brazil. In 2014, Thayane followed closely the process of Brazil's Internet Bill of Rights during her period as coordinator of the National Executive of Social Communications Students. Her research backgroud comprises the areas of public security, open government, civic technologies, feminist government, popular communications, democratization of communications and human rights. Currently, she is a researcher at the Democracy and Technology team of the Institute for Technology and Society.

İbrahim Hatipoğlu received his Masters in Public Administration (MPA) from the Bursa Uludağ University in 2014. He is a research assistant and also PhD student at the Department of Political Science and Public Administration at Bursa Uludag University, Turkey. His research interests are social media in the public sector, migration management and public management reform. He is a co-editor of "Sub-National Democracy and Politics through Social Media".

Naci Karkin is a full professor in the department of Political Science and Public Administration at Faculty of Economics and Administrative Sciences of Pamuk-kale University, Turkiye. He was also a visiting fellow in the ICT section of the Engineering, Systems and Services department at Faculty of Technology, Policy and Management of Delft University of Technology, The Netherlands. His main research interests include e-government, social media, e-participation, Web 2.0, and public policy making. Dr. Karkin has many publications in the ICTs and public administration field, book chapters of international publishing houses like Springer, Nova, IGI Global and has articles published in top information science journals like IJPADA, GIQ, IJIM, APJPA and ITD.

Olga Kolotouchkina has a PhD in Communication Sciences. Professor of Audiovisual Communication and Brand Management at CEU San Pablo University. Visiting researcher at Waseda University of Tokyo and the University of Toronto. Member of the International Place Branding Association and the American Advertising Academy. More than 18 years of professional experience as strategic planner in the field of advertising, branding and business communication at J. Walter Thompson, Saffron Brand Consultants and STUFF design consultants. Strategic consultant of the Smart Cities research project at Telefonica Spain and National Planning Strategy of the Sultanate of Oman.

Marco Konopacki is a Doctoral Research Fellow at The Governance Lab at New York University and the Institute for Technology and Society (ITS). He holds a Master's Degree from the Federal University of Paraná (UFPR) in Political Science. From 2014 to 2016, Marco was an advisor in the Secretary of Legislative Affairs of the Ministry of Justice of Brazil, where he coordinated the public debate on the regulation of Brazil's Internet Bill of Rights. Since 2019 Marco is a Fulbright H. Hubert Humphrey Fellow for tech policy and e-government at Syracuse University with his professional affiliation conducted at the House of Representatives of the United States of America.

Mehmet Fürkan Korkmaz received his MA in Political Science and Public Administration from the Bursa Uludağ University in 2016. He is a research assistant and also PhD student at the Department of Political Science and Public Administration at Bursa Uludag University, Turkey. His research interests are social media in public sector, public policy and administrative reform.

Mariana Lameiras is a Senior Academic Fellow at the United Nations University Operating Unit on Policy-Driven Electronic Governance (UNU-EGOV) and researcher of the Communication and Society Research Centre (CSRC) of the University of Minho (Braga-Portugal) in the field of Political Economy of Communication and Communication Policy, with particular interest in governance and media regulation. Currently, her research areas are e-Government, (e-Government) monitoring and assessment, e-Participation and Digital Media. She is national correspondent of the European Audiovisual Observatory collaborator (EAO) and the Institute for Information Law (University of Amsterdam), developing collective studies and writing articles on a permanent basis for IRIS - Legal Observations of the EAO and Merlin database.

João Martins is an Academic Fellow at UNU-EGOV. He is also an Invited Professor and a PhD candidate at the Department of Economics of University of Minho. As a Professor, João has lectured the courses of Public Finance, International Monetary Economics and Principles of Macroeconomics. His main research interests are in the areas of Public Economics, Political Economy, and e-Governance. He has been a member of the programme committee of several conferences in the e-Government field, such as ICEGOV, DTGS, and EGOSE.

John G. McNutt is Professor in the Biden School of Public Policy and Administration at the University of Delaware. Dr. McNutt is a specialist in the application of high technology to political and social engagement. His work focuses on the role of technology in lobbying, e-government and e-democracy, political campaigning and deliberation, organizing and other forms of political participation. He has conducted research on professional associations, child advocacy groups, consumer and environmental protection groups, social action organizations and legislative bodies. His most recent work looks at Web 2.0 political change technology and e-government and fiscal transparency. Dr. McNutt has co-edited or co-authored seven books and many journal articles, book chapters and other publications. He regularly presents at National and International conferences and sits on several editorial boards.

Uroš Pinterič is full professor of political science at the Alexander Dubček University in Trenčin, Slovakia and Jean Monnet Chair at the Faculty of Organisation Studies in Novo mesto, Slovenia. In the past he was (among other positions) a researcher at the Charles University, Czech Republic. His research interest is rather diverse and combines questions of societal identity, use of the information and communication technologies and regional development.

Rodrigo Sandoval-Almazan is Associate Professor at the Political Sciences and Social Sciences Department in the Autonomous State University of Mexico, in Toluca City. Dr. Sandoval-Almazan is the author or co-author of articles in Government Information Quarterly, Information Polity, Electronic Journal of Electronic Government; Journal of Information Technology for Development; Journal of Organizational Computing and Electronic Commerce; International Journal of E-Politics IJEP. His research interests include artificial intelligence, social media in government. public innovation, digital government, and open government. Professor Sandoval-Almazan has a Bachelor's Degree in Political Science and Public Administration, a Masters in Management focused on Marketing, and a Ph.D. in Management with Information Systems.

Ecem Buse Sevinç Çubuk is a research assistant in the Department of Public Administration at Faculty of Business Administration of Aydin Adnan Menderes University, Turkiye. She graduated from Bilkent University, Turkiye with a Bachelor of International Relations in 2006. As a Ph.D. candidate in the program of Political Science and Public Administration at the Graduate School of Social Sciences of Pamukkale University, Turkiye, she is an ABD graduate student currently writing a dissertation concerned with the relation between ICTs-led innovation in the public sector and the public value(s) creation. She is also a guest researcher in the ICT section of the Engineering, Systems, and Services department at the Faculty of Technology, Policy and Management of the Delft University of Technology, The Netherlands. Her interests include the digital transformation of public services, policy, and practice of innovation management, big data, and smart cities.

Mehmet Zahid Sobaci is a Full Professor in the Department of Political Science and Public Administration at Bursa Uludağ University, Turkey. His research interests are social media in politics; public policy, and public management reform. He has authored, edited, and co-edited books, most recently "Sub-National Democracy and Politics through Social Media", "Social Media and Local Governments: Theory and Practice", "Administrative Reform and Policy Transfer: The Diffusion of the New Public Management (in Turkish), and "Public Policy: Theory and Practice" (in Turkish). He is also the author of many journal articles and book chapters published in national and international journals and books.

António Tavares holds a Ph.D. in Public Administration from Florida State University. He is associate professor of the School of Economics and Management and member of the Research Center in Political Science at the University of Minho, Portugal. He is also adjunct associate professor at the United Nations University, Operating Unit on Policy-Driven Electronic Governance (UNU-EGOV) in Guimarães, Portugal. He is co-editor of the Urban Affairs Review, the journal affiliated with the Urban Politics Section of the American Political Science Association. His research interests comprise topics in the fields of local government and urban politics, including territorial reforms, regional governance, service delivery, and political and civic engagement. His research has been published in a wide range of journals in political science, public administration and urban affairs, including Government Information Quarterly, Journal of Urban Affairs, Local Government Studies, Policy Studies Journal, International Review of Administrative Sciences, and Public Management Review.

David Valle-Cruz is Professor at the Computing Engineering Department in the Autonomous University of the State of Mexico, in Toluca City. Dr. David Valle-Cruz is the author or co-author of articles in Information Polity, First Monday, International Journal of Public Sector Management, International Journal of Public Administration in the Digital Age, and Digital Government: Research and Practice. His research is related to Applied Artificial Intelligence and Emerging Technologies in Government.

Nilay Yavuz is an Associate Professor of Political Science and Public Administration at Middle East Technical University in Turkey. She received her PhD from University of Illinois at Chicago. Her research interests include digital government, individual technology adoption, public participation, public policy analysis, organizational behavior, and public transportation. She has published journal articles in Government Information Quarterly, Urban Studies, International Journal of Public Administration, IJPADA, and Transportation Research Record, and authored and co-authored several book chapters. She has recently completed a research project on the municipal service quality and public participation relationship, funded by the Scientific and Technological Research Council of Turkey.

Index

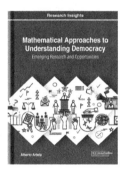
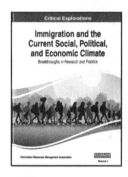
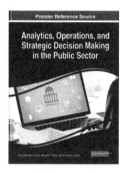

Ensure Quality Research is Introduced to the Academic Community

Become an IGI Global Reviewer for Authored Book Projects

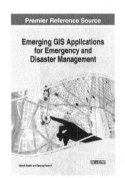

Premier Reference Source

Emerging GIS Applications for Emergency and Disaster Management

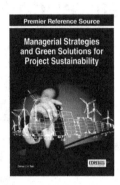

Premier Reference Source

Managerial Strategies and Green Solutions for Project Sustainability

Premier Reference Source

Comparative Approaches to Using R and Python for Statistical Data Analysis

Premier Reference Source

Solutions for High-Touch Communications in a High-Tech World

The overall success of an authored book project is dependent on quality and timely reviews.

In this competitive age of scholarly publishing, constructive and timely feedback significantly expedites the turnaround time of manuscripts from submission to acceptance, allowing the publication and discovery of forward-thinking research at a much more expeditious rate. Several IGI Global authored book projects are currently seeking highly-qualified experts in the field to fill vacancies on their respective editorial review boards:

Applications and Inquiries may be sent to:
development@igi-global.com

Applicants must have a doctorate (or an equivalent degree) as well as publishing and reviewing experience. Reviewers are asked to complete the open-ended evaluation questions with as much detail as possible in a timely, collegial, and constructive manner. All reviewers' tenures run for one-year terms on the editorial review boards and are expected to complete at least three reviews per term. Upon successful completion of this term, reviewers can be considered for an additional term.

If you have a colleague that may be interested in this opportunity, we encourage you to share this information with them.

Printed in the United States
By Bookmasters